The Wykeham Science Series

General Editors:

PROFESSOR SIR NEVILL MOTT, F.R.S.

Emeritus Cavendish Professor of Physics
University of Cambridge

G.R. NOAKES

Formerly Senior Physics Master
Uppingham School

The Author:

A.P. CRACKNELL is professor of theoretical physics at the University of Dundee, and has previously been a lecturer at the Universities of Singapore and Essex. He has written numerous research papers and seven books on various aspects of solid-state physics.

The Schoolmaster:

J.L. CLARK is Principal Teacher of physics at Menzieshill High School, Dundee.

ULTRASONICS

A.P. CRACKNELL
University of Dundee

WYKEHAM PUBLICATIONS (LONDON) LTD
(A member of the Taylor and Francis Group)
LONDON and BASINGSTOKE 1980

First published 1980 by Wykeham Publication (London) Ltd.

Printed in Great Britain by Alden Press, Oxford, London and Northampton

British Library Cataloguing in Publication Data

Cracknell, Arthur Philip
Ultrasonics. – (The Wykeham science series; 55).
1. Ultrasonic waves
I. Title
534.5′5 QC244

ISBN 0-85109-770-7

Preface

During the last two or three decades, ultrasonic techniques have been developed so much that ultrasound can now be regarded as a significant branch of physics which has numerous applications in everyday life. This small book attempts to describe, in outline, the physics of the generation, propagation, attenuation, and detection of ultrasound (Chapters 2 to 4) and to indicate the principal ways in which physics is exploited for the benefit of mankind (Chapters 5 to 10).

This is intended to be an introductory text; it grew out of an article which my wife and I wrote for *Contemporary Physics* in 1976. At the end of the book there is a list of suggested further reading for anyone who wishes to study some particular part of the subject in more detail.

In the preparation of the present book I have been assisted by Mr J. L. Clark, Principal Teacher of Physics at Menzieshill High School in Dundee. In addition to making constructive criticism of the whole manuscript, Mr Clark also prepared the first draft of parts of chapter 1 for me. I am also grateful to the many authors, editors, industrialists, and publishers who have granted permission for the reproduction of their copyright material in tables or illustrations; the sources of these are indicated *in situ*. I am also grateful to Miss M.M. Benstead for re-drawing the diagrams for figures 5.5 and 7.30 and to Mr J.Q. Summers for some help in obtaining references on ophthalmology.

Dundee

A.P. Cracknell
July 1979

Contents

1. Introduction

1.1. *Ultrasound*

By the term *ultrasound* we mean vibrations of a material medium which are similar to sound waves, but which have frequencies that are too high to be detected by an average human ear. The study and applications of these vibrations are called *ultrasonics*. In early days the term 'supersonics' was also used to include what we now describe as ultrasonics, but this term is now used only for *speeds* greater than the speed of 'ordinary' sound.

The upper frequency limit of human hearing varies from about 10 kHz to about 18 kHz. For any given person this threshold frequency decreases with increasing age. Towards the end of the nineteenth century Galton described the use of a small ultrasonic whistle to investigate the threshold frequency of hearing for humans.*

> On testing different persons, I found there was a remarkable falling off in the power of hearing high notes as age advanced. The persons themselves were quite unconscious of their deficiency so long as their sense of hearing low notes remained unimpaired. It is an only too amusing experiment to test a party of persons of various ages, including some rather elderly and self-satisfied personages. They are indignant at being thought deficient in the power of hearing, yet the experiment quickly shows that they are absolutely deaf to shrill notes which the younger persons hear acutely, and they commonly betray much dislike to the discovery. Every one has his limit, and the limit at which sounds become too shrill to be audible to any particular person can be rapidly determined by this little instrument.

Galton also studied the upper threshold frequency of hearing of various animals. There is no reason to suppose that this should be the same as for man (see fig. 1.1). The higher threshold frequency of birds is exploited in the construction of ultrasonic bird-scarers that are inaudible to man, while bats are known to use ultrasonic pulse-echo techniques 'to see in the dark' (see Chapter 6).

The spectrum of acoustic vibrations is illustrated in fig. 1.2. In many

*F. Galton, *Inquiries into Human Faculty and Development* (Macmillan, 1883).

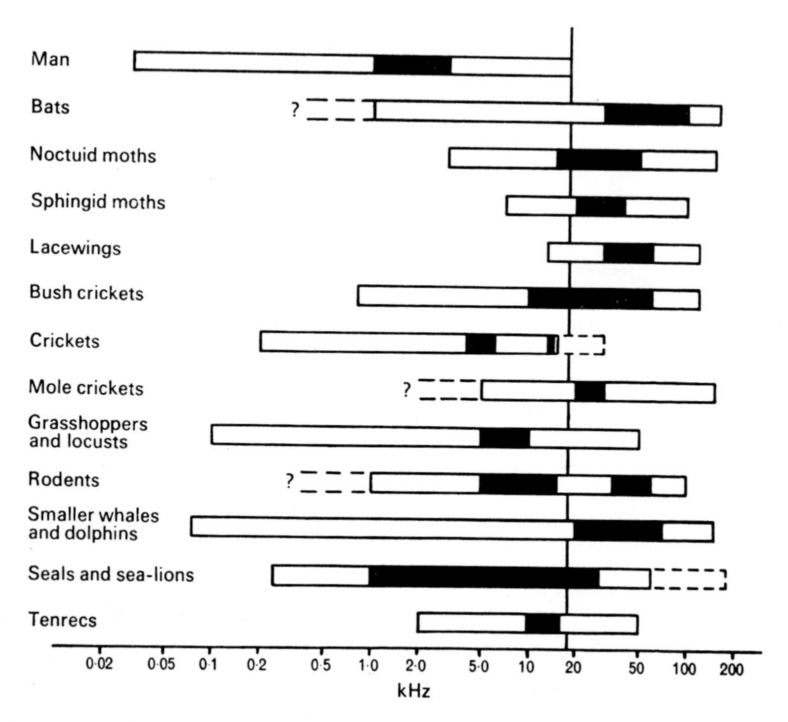

Fig. 1.1. The frequency range of hearing in man and approximate ranges for some groups of animals. The dark regions show the most sensitive frequencies. (From G. Sales and D. Pye, *Ultrasonic Communication by Animals* (Chapman and Hall, London, 1974)).

of the non-scientific and non-medical applications of ultrasound there would be no advantage in using very high frequencies; often the main consideration is simply to be out of the audible range so as to avoid discomfort to the workers, although the variation of the attenuation with frequency may also influence the choice of frequency that is used. In some of the scientific applications, however, it is important to use the highest available frequencies. At the present time any upper limit to ultrasonic frequencies is set by practical considerations rather than by theory. The term *microwave acoustics* is often used to describe the study of mechanical vibrations with frequencies which are so very high that they correspond to frequencies in the microwave range of the electromagnetic spectrum, say $\geqslant 10$ GHz, the vibrations themselves at these high frequencies being referred to as *microsound* or *praetersound* (fig. 1.2).

The early work on the upper limit of audibility of sound was performed using ultrasonic generators in the form of whistles (Galton whistles) which

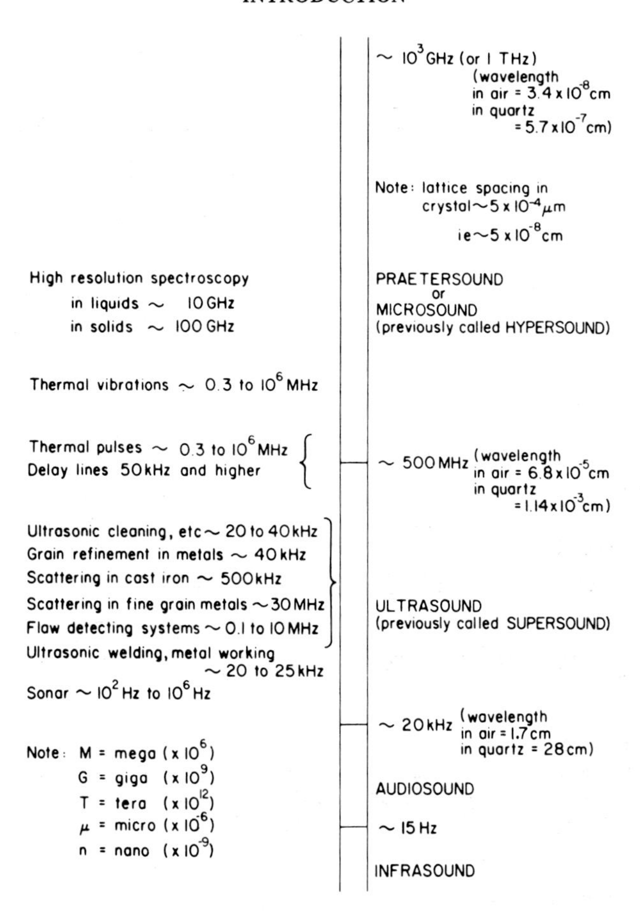

Fig. 1.2. Acoustic frequency scale (From R.W.B. Stephens, *Ultrasonics International 1975 Conference Proceedings*, p. 9, (IPC Science and Technology Press, Guildford, 1975)).

had clearly been developed from organ pipes (see Chapter 4). The detection of ultrasound in the early days was performed using sensitive flames. The development of modern ultrasonic generators and detectors, and the whole modern technology of ultrasonics, has resulted mainly, though not exclusively, from the exploitation of piezoelectricity (see Section 4.2), backed up by modern electronic techniques. Piezoelectricity was discovered by the Curie brothers in 1880. In this phenomenon, some materials when they are under some external mechanical stress develop an internal electric field, with charges of opposite sign appearing on opposite surfaces. In the following year Lippmann predicted the inverse effect, namely the appearance of a mechanical deformation of certain materials in external electric fields. However, it was not until after the 1914–18 War that piezoelectricity and its inverse effect were first successfully applied in the detection and generation of ultrasound (see Section 5.1). The original application was to the detection of enemy submarines.

In this book we shall first be concerned with the physics of the production, propagation, attenuation, and detection of ultrasound (Chapters 2 to 4). After that we shall describe the general principles of the main present-day applications of ultrasonics in biology, medicine, technology and pure research in physical science. The wide range of these applications of ultrasound at the present time will become apparent as one reads through the book.

1.2. *Wave motion*

In this section we shall consider the propagation of waves in a material medium; this discussion will be equally relevant to ultrasound and to audible sound because, at this stage, the value of the frequency will not be restricted. In any event, the minimum frequency that may be classed as ultrasonic is neither fixed very precisely, nor is it a frequency at which any distinct change of wave properties suddenly occurs.

Let us consider a longitudinal progressive wave in which the particles of the medium move backwards and forwards along a *line* which is the same line as that in which the wave is travelling. It is, in effect, a one-dimensional wave. It may be referred to as a compressional wave, and an instantaneous picture of it is as fig. 1.3(*a*) and (*b*). The graph of the instantaneous value of the longitudinal displacement as a function of x is as shown in fig. 1.3(*c*), which should be treated with a little caution, lest it should obscure the fact that the wave is actually longitudinal.* In fig. 1.3(*c*) we have assumed, as is usual in a first approach, that the displacements are sinusoidal. This need not necessarily be the case but, for two reasons, sinusoidal, or simple harmonic, waves are the most important ones in practice. First, ultrasound is usually produced by the conversion of sinusoidal electromagnetic oscillations, of a single frequency, into ultrasonic waves. Secondly, any wave can always be Fourier-analysed into components which are pure sinusoidal waves.

Both fig. 1.3(*b*) and (*c*) represent instantaneous 'snapshots' of the medium taken at one instant of time t. At some later instant the particles

*Transverse ultrasonic waves may, of course, exist in certain circumstances, see Section 2.1.

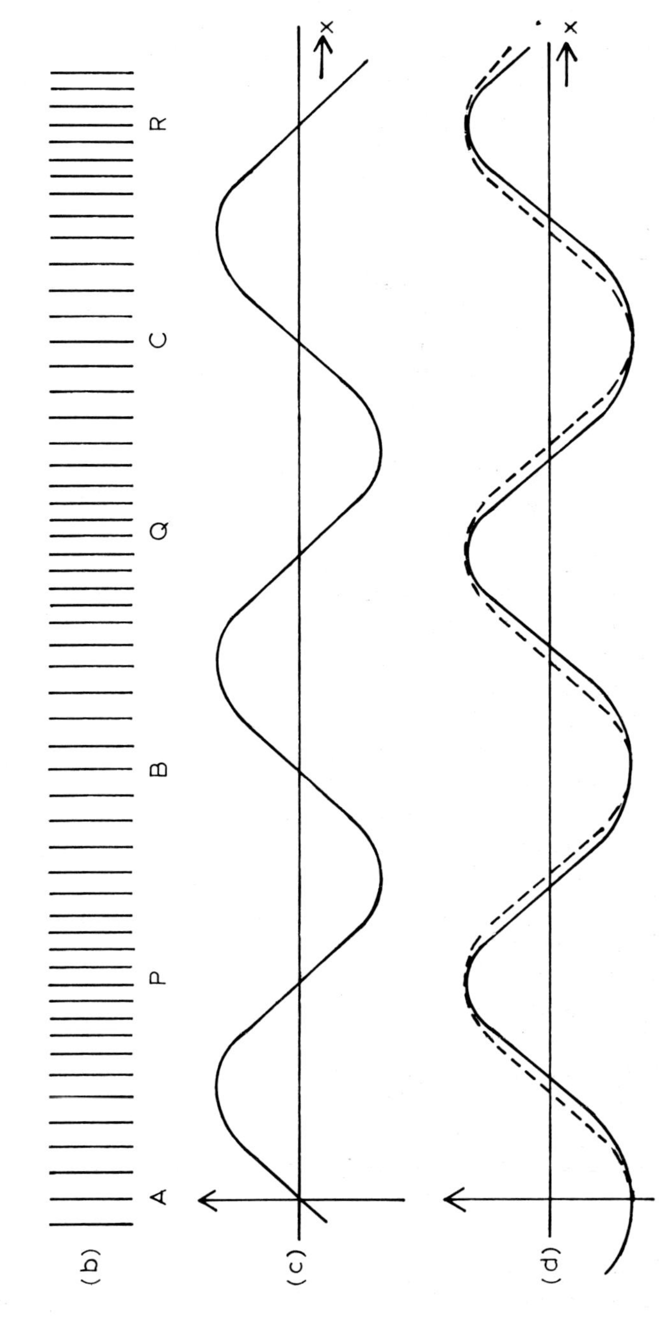

Fig. 1.3. Schematic illustration of longitudinal wave, (*a*) equally spaced planes in the absence of wave, (*b*) displaced positions of same planes at one instant during passage of wave, A, B, C are regions of rarefaction, P, Q, R are regions of compression (*c*) graph of instantaneous displacements, (*d*) graph of excess pressure (continuous curve)—the broken curve represents the excess pressure as a function of equilibrium positions of the particles.

of the medium will have rearranged themselves, while the compressions and rarefactions will have travelled a certain distance from, say, left to right, in the diagram. The individual particles of the medium are oscillating about their equilibrium positions without there being any net movement of the medium as a whole—one does not, for example, experience a wind blowing when listening to an orchestra or a choir!

The pressure in a region of compression is greater than the pressure would be in the medium in the absence of the wave. Thus one could plot a graph of the excess pressure $p(x, t)$, at a fixed time, as a function of x along the path of the wave (see fig. 1.3(*d*)).

The *displacement* of a particle is its distance from its equilibrium position; this is a function both of the equilibrium position x and of the time t and so can be written $u(x, t)$. For a sinusoidal wave as in fig. 1.3(*c*), each particle participating in the wave motion is vibrating with simple harmonic motion about its own equilibrium position. The amplitude a of the wave is the maximum value of the displacement, and the *wavelength* λ is the distance between any two consecutive particles which are vibrating in phase.

The expression for the particle displacement $u(x, t)$, as a function of x and t, can be written

$$u(x, t) = a \sin\left(\frac{2\pi t}{T} - \frac{2\pi x}{\lambda}\right), \tag{1.1}$$

where λ is the wavelength. The time T required for one complete cycle of a particle's motion is called the *period* of the oscillation. This is related to the *frequency* ν of the oscillation by

$$\nu = 1/T. \tag{1.2}$$

The angular frequency ω is defined by

$$\omega = 2\pi\nu. \tag{1.3}$$

Equation (1.1) can be rewritten in a number of ways. In terms of the frequency $\nu(= 1/T)$, instead of the period T, it becomes

$$u(x, t) = a \sin\left(2\pi\nu t - \frac{2\pi x}{\lambda}\right) = a \sin 2\pi\left(\nu t - \frac{x}{\lambda}\right). \tag{1.4}$$

c, the speed of the wave, is related to λ and ν since during a single period of a *particle's* motion the travelling *wave* will advance by one wavelength. Thus

$$c = \lambda/T = \nu\lambda. \tag{1.5}$$

In Table 1.1. we give values of λ, for various frequencies ν, for

Table 1.1. Some examples of ultrasonic wavelengths.

Frequency	Wavelength		
	(for $c = 1000\ \mathrm{m\,s^{-1}}$)	(for $c = 3000\ \mathrm{m\,s^{-1}}$)	(for electromagnetic radiation)
20 kHz = 2×10^4 Hz	5 cm	15 cm	1.5×10^4 m
100 kHz = 10^5 Hz	1 cm	3 cm	3×10^3 m
1 MHz = 10^6 Hz	1 mm	3 mm	300 m
50 MHz = 5×10^7 Hz	20 μm	60 μm	6 m
1 GHz = 10^9 Hz	1 μm	3 μm	30 cm

ultrasonic speeds of 1000 m s^{-1} and 3000 m s^{-1}, which are typical speeds in a liquid and in a solid (see Table 2.1). For comparison we also give the corresponding wavelengths for electromagnetic radiation of the same frequencies.

1.3. *Energy, momentum and pressure in an ultrasonic wave*

The *intensity* of a wave is the rate at which energy is transferred across a unit area normal to the direction of propagation of the wave, that is the power transferred per unit area. Appropriate units are therefore J s^{-1} m^{-2} = W m^{-2}.

Consider one particle, of mass m, of the medium through which such a wave is travelling. The particle's instantaneous velocity is found by differentiating eqn. (1.4)

$$v(x, t) = \frac{\partial u(x, t)}{\partial t} = a2\pi\nu \cos 2\pi\left(\nu t - \frac{x}{\lambda}\right). \tag{1.6}$$

The maximum value of this velocity is therefore

$$v_{\text{max}} = a2\pi\nu = a\omega. \tag{1.7}$$

The energy E of the particle will, in general, be partly potential energy E_p and partly kinetic energy E_k, so

$$E = E_p + E_k. \tag{1.8}$$

Consider the instant when the particle is passing through its equilibrium position. The potential energy is zero and its velocity has its maximum value. The total energy then is therefore equal to the kinetic energy, thus

$$E = \tfrac{1}{2}mv_{\text{max}}^2 = \tfrac{1}{2}ma^2\omega^2. \tag{1.9}$$

If the medium has a density ρ, then a volume V of the medium will have a mass ρV and will contain $\rho V/m$ particles. The total energy contained in a volume V is therefore given by

$$\text{total energy in volume } V = \tfrac{1}{2}ma^2\omega^2\frac{\rho V}{m} = \tfrac{1}{2}a^2\omega^2\rho V. \tag{1.10}$$

The total energy of all the particles in unit volume is called the *energy density* E and is therefore given by

$$E = \tfrac{1}{2}a^2\omega^2\rho. \tag{1.11}$$

This energy is being transferred by the wave at a speed c. Consider a cuboid of length l and cross-sectional area A, the energy in which is to be transferred across plane X (fig. 1.4). The energy contained in the cuboid,

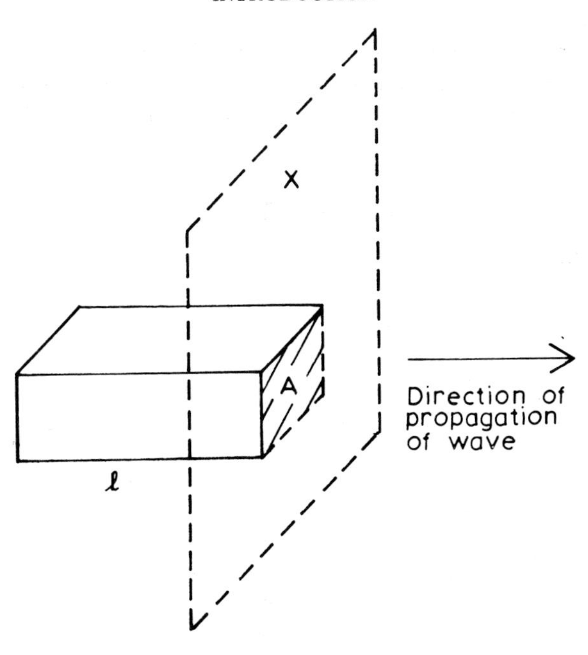

Fig. 1.4.

which is $\frac{1}{2}a^2\omega^2\rho lA$, will cross the plane X in a time given by l/c. Hence the power P transferred is given by

$$P = \frac{\frac{1}{2}a^2\omega^2\rho lAc}{l} = \frac{1}{2}a^2\omega^2\rho Ac. \tag{1.12}$$

The power transferred across unit area, that is the intensity I, is therefore given by

$$I = \frac{1}{2}a^2\omega^2\rho c. \tag{1.13}$$

Having seen that a travelling ultrasonic wave transports energy, it is natural to expect that it also carries momentum. It is also natural to expect that if such a wave impinges on a surface and is reflected or absorbed by the surface, there will be a pressure exerted on that surface. Such a pressure can, indeed, be detected experimentally. One might call this pressure the *radiation pressure* but, in fact, there are two different definitions of radiation pressure. One, the *Rayleigh radiation pressure*, is taken to be the difference between the time average of the pressure at a point in a fluid through which a beam of sound, or ultrasound, passes and the pressure which would have existed in a fluid of the *same mean density* at rest. The other, the *Langevin radiation pressure*, is defined as the difference between the pressure at a wall and the pressure in the medium, at rest, behind the wall. Much as one would like to treat the transport of momentum and the phenomenon of radiation pressure quantitatively, the theory in both cases is rather too complicated for this book (see Further Reading).

2. The propagation of ultrasound

2.1. *Propagation in an isotropic medium*

Since ultrasound is, essentially, of the same nature as audible sound, the theory of its propagation is similar to that for audible sound. In this section we shall restrict ourselves to isotropic media. For the present purpose, 'isotropic' means that the magnitude of the ultrasonic velocity is the same for all possible directions of propagation in the medium. Isotropic media include gases, liquids, non-crystalline solids and polycrystalline solids. It should be noted that whereas cubic crystals are isotropic in some of their properties they are not isotropic as far as the propagation of sound or ultrasound is concerned. We shall consider the propagation of ultrasound in crystals in Chapter 9.

The three conventional elastic moduli for an isotropic material are the Young modulus E, the bulk modulus K, and the shear modulus G.

The Young modulus is defined in terms of the stretching of long thin wires or rods (fig. 2.1(*a*)) with

$$\text{stress} = \frac{\text{stretching tension}}{\text{area of cross-section}} = \frac{F}{A} \tag{2.1}$$

and

$$\text{strain} = \frac{\text{increase in length}}{\text{original length}} = \frac{\delta L}{L} \tag{2.2}$$

so that

$$E = \frac{\text{stress}}{\text{strain}} = \frac{F/A}{\delta L/L} = \frac{FL}{A\delta L}. \tag{2.3}$$

The bulk modulus is concerned with the compression of a material under a (uniform) hydrostatic pressure p (fig. 2.1(*b*)) when

$$\text{stress} = \text{pressure} = p \tag{2.4}$$

and

$$\text{strain} = \frac{\text{increase in volume}}{\text{original volume}} = -\frac{\delta V}{V} \tag{2.5}$$

so that

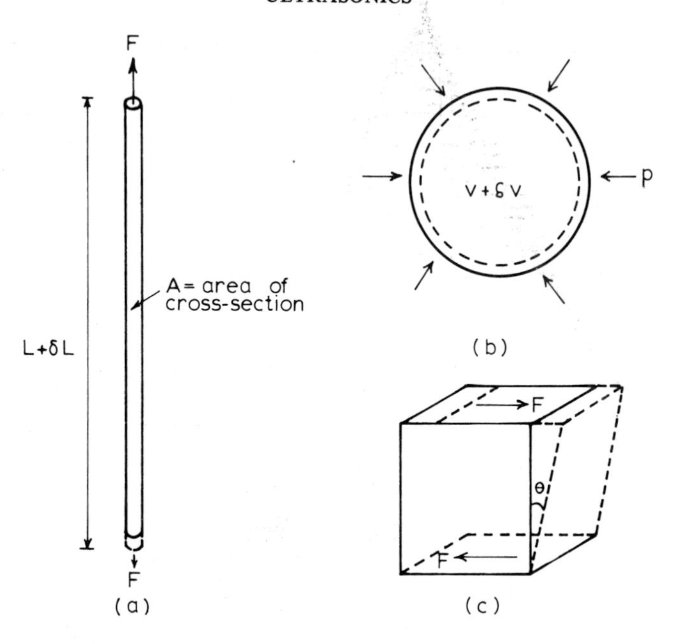

Fig. 2.1. Diagrams to illustrate the definitions of (*a*) the Young modulus, (*b*) the bulk modulus, and (*c*) the shear modulus.

$$K = \frac{\text{stress}}{\text{strain}} = \frac{p}{-\delta V/V} = -\frac{pV}{\delta V}. \tag{2.6}$$

Finally, for the shear modulus (fig. 2.1(*c*))

$$\text{stress} = \frac{\text{shearing force}}{\text{area}} = \frac{F}{A} \tag{2.7}$$

and

$$\text{strain} = \text{angle of shear} = \theta \tag{2.8}$$

so that

$$G = \frac{F/A}{\theta} = \frac{F}{A\theta}. \tag{2.9}$$

We shall assume that a medium is behaving in a linear manner so that moduli of elasticity are constants for any given material; that is, the stress is assumed to be sufficiently small that Hooke's law is obeyed.

For a solid these three moduli of elasticity are usually considered independent although it is possible to obtain relationships among them by introducing Poisson's ratio. Which particular modulus is relevant will depend on the system in which the waves are propagating; both longitudinal (compressional) waves and transverse (shear) waves may occur. The Young modulus applies only to solids. If a *steady* stress is applied to a fluid, the fluid will flow so as to eliminate the consequent strain; the relevant parameter is the viscosity which is concerned with the rate at which the shear strain is eliminated. However, the strain in a fluid does not disappear instantaneously as soon as the stress is applied and so the shear modulus applies to fluids as well as solids. If a rapidly oscillating shear stress is applied to a liquid or gas, there will be a corresponding oscillating shear strain. It should, therefore, be possible to launch transverse (shear) ultrasonic waves into a liquid, or even into a gas, although such waves will be very heavily attenuated. The bulk modulus K is relevant for gases, liquids and solids.

We shall now consider longitudinal waves travelling along a rod, with area of cross section A and material of Young modulus E (fig. 2.2). PQ is a small element of this rod, the equilibrium positions of its ends being at x and $(x + \delta x)$. Thus δx is the equilibrium length of this element in the absence of strain. Suppose also that the displacements of the ends of the rods are u and $(u + \delta u)$, respectively. u is a function of x and t and could be written as $u(x, t)$. Then the coordinates of P′ and Q′, the displaced positions of the ends of the element, are

$$P':x + u$$

$$Q':x + \delta x + u + \delta u$$

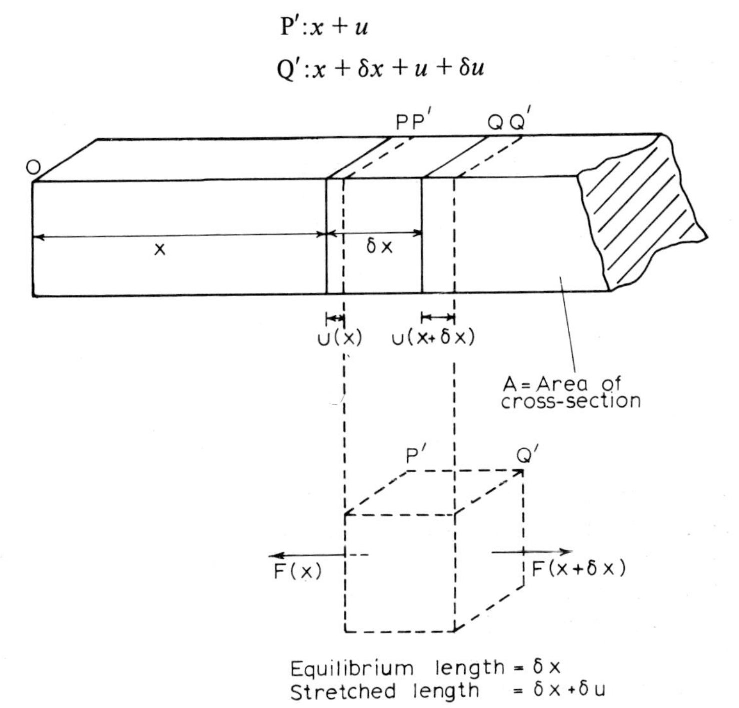

Fig. 2.2. Displacement of a small element of a rod transmitting longitudinal vibrations.

and the length of the element during the vibration is

$$(x+\delta x+u+\delta u)-(x+u) = (\delta x+\delta u). \tag{2.10}$$

Since δx is the equilibrium length of the element, the increase in its length is δu and so the strain is $\delta u/\delta x$. Or, if we regard the strain as varying continuously along the length of the rod, we may represent the strain at each point along the rod by $(\partial u/\partial x)$. In defining the Young modulus (see fig. 2.1(*a*)), each point of the rod or wire is at rest, and both the stress and the strain are independent of position in the rod or wire. When an ultrasonic wave is travelling along the rod, the values of both stress and strain vary from point to point along it. But, we can still use the definition of the Young modulus (eqn. (2.3)) as a local relationship between the stress at any given point in the rod and the strain at the same point. Thus

$$E = \frac{F(x)/A}{(\partial u/\partial x)}, \tag{2.11}$$

and so

$$F(x) = EA\frac{\partial u}{\partial x}. \tag{2.12}$$

It is the difference between the forces $F(x+\delta x)$ and $F(x)$ at the ends of the small element PQ in fig. 2.2 that provides the restoring force on the element. Substituting this into eqn. (2.1) gives

$$EA\left(\frac{\partial u}{\partial x}\right)_{\mathrm{Q}'} - EA\left(\frac{\partial u}{\partial x}\right)_{\mathrm{P}'} = A\delta x\rho\frac{\partial^2 u}{\partial t^2} \tag{2.13}$$

where ρ is the density. But,

$$\frac{\left(\frac{\partial u}{\partial x}\right)_{\mathrm{Q}'} - \left(\frac{\partial u}{\partial x}\right)_{\mathrm{P}'}}{\delta x} \simeq \frac{\partial^2 u}{\partial x^2} \tag{2.14}$$

so that eqn. (2.13) becomes

$$E\frac{\partial^2 u}{\partial x^2} = \rho\frac{\partial^2 u}{\partial t^2}, \tag{2.15}$$

and the general differential equation for the propagation of waves in one dimension is

$$\frac{\partial^2 u}{\partial x^2} = \frac{1}{c^2}\frac{\partial^2 u}{\partial t^2} \tag{2.16}$$

where c is the speed of the waves. Equation (2.15), which applies to longitudinal elastic waves travelling along a rod, can be written in this form by taking $c = \sqrt{(E/\rho)}$. The solutions of eqn. (2.16) that will give expressions for the displacement $u(x, t)$ at a point x along the bar at a time t may be written as

$$u(x, t) = a \exp [\mathrm{i}(\omega t - kx)] \tag{2.17}$$

and

$$u(x, t) = a \exp [\mathrm{i}(\omega t + kx)]. \tag{2.18}$$

In these equations we use this exponential form, which is equivalent to the form used in eqn. (1.4), and we also use the quantity k, often called the *wave vector*, which is related to the wavelength by

$$k = 2\pi/\lambda. \tag{2.19}$$

The speed c can also be expressed in terms of ω and k by using eqns. (1.3) and (1.5)

$$c = \nu\lambda = \frac{\omega}{2\pi}\frac{2\pi}{k} = \frac{\omega}{k}. \tag{2.20}$$

This argument can be generalized to three-dimensional systems such as regions of gaseous, liquid, or solid media in which a wave is being propagated in an arbitrary direction. By considering the forces on a small element of the medium it is possible to show that

$$\frac{\partial^2 u}{\partial x^2} + \frac{\partial^2 u}{\partial y^2} + \frac{\partial^2 u}{\partial z^2} = \frac{1}{c^2}\frac{\partial^2 u}{\partial t^2} \tag{2.21}$$

where

$$c = \sqrt{\left(\frac{\mathscr{E}}{\rho}\right)} \tag{2.22}$$

and $\mathscr{E}$ is the appropriate elastic modulus. There are many different possible types of solution $u(x, y, z, t)$ to eqn. (2.21); these correspond to plane waves, spherical waves, cylindrical waves and other more complicated patterns. The plane-wave solutions take the form

$$u(x, y, z, t) = a \exp\{i[\omega t \pm (k_1 x + k_2 y + k_3 z)]\}. \tag{2.23}$$

The plane wave is characterized by the wave vector $\mathbf{k}$ which has components k_1, k_2, and k_3 parallel to the x, y, and z axes, respectively. The direction of $\mathbf{k}$ is the direction in which the plane wave is travelling (see fig. 2.3), and its magnitude $|\mathbf{k}|$, or k, is related to the wavelength by

$$|\mathbf{k}| = \frac{2\pi}{\lambda}. \tag{2.24}$$

For any plane that is perpendicular to $\mathbf{k}$, at any given time, the displacement $u(x, y, z, t)$ will be the same all over the plane. At a later instant the value of $u(x, y, z, t)$ on that plane will have altered but it will still have a

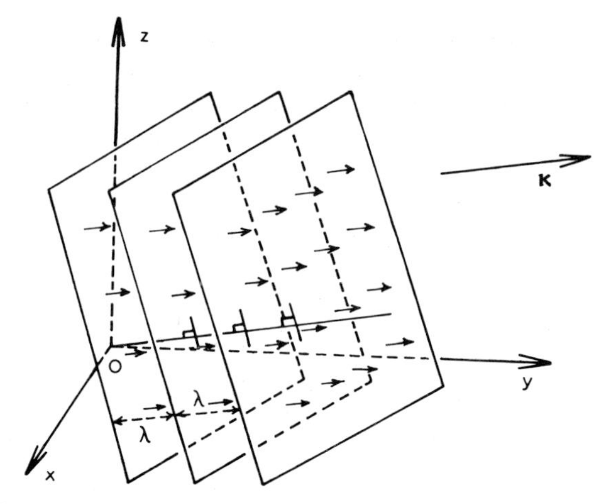

Fig. 2.3 Diagram to represent plane longitudinal wave. **k** is along the direction of propagation. The planes, which are shown as separated from one another by λ and as normal to **k**, represent successive wavefronts. The displacements, represented by the small arrows, are in phase on successive wavefronts; in this diagram for longitudinal waves the displacements are parallel to **k** and normal to the wavefronts.

common value all over the plane. If we suppose that in fig. 2.3 we have chosen the planes for an instant at which the displacements have their maximum values, these planes will be separated by λ and we can regard these planes of maximum displacement as the wavefronts. For a line parallel to **k**, and thus perpendicular to the plane wavefronts, the form of $u(x, y, z, t)$ as a function of distance along this line will be a sine (or cosine) wave travelling with velocity c. This general discussion of plane waves is relevant both to longitudinal, or compressional, waves for which $u(x, y, z, t)$ is parallel to **k** and also to transverse, or shear, waves for which $u(x, y, z, t)$ will be perpendicular to **k**.

The modulus $\mathscr{E}$ that is used in eqn. (2.22) depends on the type of wave considered. Compressional waves involve longitudinal vibrations like those of fig. 2.2, and shear waves involve transverse vibrations. In a liquid or gas, shear waves would be very heavily attenuated so that they can only be propagated over very short distances. Therefore only longitudinal (compressional) waves are of interest in a liquid or gas, when $\mathscr{E}$ will be the bulk modulus. It is the *adiabatic* bulk modulus K_a which is related to the *isothermal* bulk modulus K_i by

$$K_a = \gamma K_i \tag{2.25}$$

where $\gamma = c_p/c_v$, the ratio of the principal specific heat capacities c_p, at constant pressure, and c_v, at constant volume, for the gas. For a gas K_i is equal to the pressure p so that $K_a = \gamma p$. We have seen that for longitudinal waves in a rod the appropriate modulus of elasticity $\mathscr{E}$ is the Young modulus E. But for longitudinal waves in a three-dimensional solid (see fig. 2.3) it is not only the Young modulus that is involved. The difference is that when a compression or rarefaction occurs in a rod, the rod is free to expand or contract laterally, but when a very wide plane propagates in a three-dimensional solid (fig. 2.3), these lateral changes are prevented by the neighbouring material. It can be shown that the longitudinal compression in the plane wave in a solid is equivalent to a uniform (hydrostatic) compression and two shear stresses, and that in this case

$$\mathscr{E} = K + \tfrac{4}{3}G. \tag{2.26}$$

For transverse (shear) waves in a solid $\mathscr{E}$ is the (adiabatic) shear modulus

$$\mathscr{E} = G. \tag{2.27}$$

The expressions for $\mathscr{E}$ for various important types of waves can be summarized:

	$\mathscr{E}$ is taken as
Solid	
longitudinal (compressional) waves – rod (c_E)	E
longitudinal (compressional) waves – bulk (c_L)	$E + \frac{4}{3}G$
transverse (shear) waves (c_S)	G
Liquid	
longitudinal (compressional) waves	K (adiabatic)
Gas	
longitudinal (compressional) waves	K (adiabatic) ($= \gamma p$).

In Table 2.1 we have collected values of c for a number of materials.

The speed of sound, or the ultrasound, is very much smaller than the speed of electromagnetic radiation, which is (in vacuum) about $3 \times 10^8 \,\mathrm{m\,s^{-1}}$. This leads to an interesting use for ultrasonic waves in radar, in television sets and in digital computers, namely to delay an electromagnetic wave. This delay could be achieved by sending the electromagnetic wave on a round trip in air or in vacuum. But to achieve a delay of, say, 1 ms would require a distance of $3 \times 10^8 \times 10^{-3}\,\mathrm{m} = 300\,\mathrm{km}$, which is a highly inconvenient, if not actually impossible path length. However, with the device illustrated in fig. 2.4 one can achieve a delay of 1 ms with a quite short path length. In such a *delay line*, an electromagnetic wave is converted by a *transducer* (see Chapter 4) into an ultrasonic wave which is launched at A into the solid medium AB. The ultrasonic wave travels from A to B and then at B it is converted back into an electromagnetic wave by a second transducer. If the ultrasonic wave travels in the solid at, say,

Table 2.1. Velocities and acoustic impedances for some materials.

(a) *Gases and liquids*

	Speed $m\,s^{-1}$	Acoustic impedance $kg\,m^{-2}\,s^{-1}$
Air	331·46	431
Carbon dioxide	259	512
Helium	971·9	173
Hydrogen	1286	116
Neon	434	391
Nitrogen	337	421
Distilled water	1482·3	$1{\cdot}48 \times 10^{6}$
Acetic acid	1173	$1{\cdot}23 \times 10^{6}$
Acetone	1190	$9{\cdot}37 \times 10^{5}$
Carbon tetrachloride	940	$1{\cdot}94 \times 10^{6}$
Ethanol	1162	$9{\cdot}17 \times 10^{5}$
Glycerol	1860	$2{\cdot}34 \times 10^{6}$
Mercury	1454	$1{\cdot}97 \times 10^{7}$

(b) *Solids*

	Speed for rod waves, c_E $m\,s^{-1}$	Speed for compressional waves, c_L $m\,s^{-1}$	Speed for shear waves, c_S $m\,s^{-1}$	Acoustic impedance (compressional waves) $kg\,m^{-2}\,s^{-1}$
Aluminium	5102	6374	3111	$1{\cdot}7 \times 10^{7}$
Beryllium	?	12890	8880	$2{\cdot}3 \times 10^{7}$
Brass	3451	4372	2100	$3{\cdot}7 \times 10^{7}$
Crown glass	5342	5660	3420	$1{\cdot}4 \times 10^{7}$
Lead	1188	2160	700	$2{\cdot}4 \times 10^{7}$
Perspex	2177	2700	1330	$3{\cdot}2 \times 10^{6}$
Sandstone	2820	2920	1840	$4{\cdot}7 \times 10^{7}$
Soft iron	5189	5957	3224	$4{\cdot}7 \times 10^{7}$
Zinc	3826	4187	2421	$3{\cdot}0 \times 10^{7}$

Note: Data for gases are for 0°C; data for liquids and solids are for 20°C. (Values extracted from G.W.C. Kaye and T.H. Laby, *Tables of Physical and Chemical Constants* 14th ed. (Longmans, 1973).)

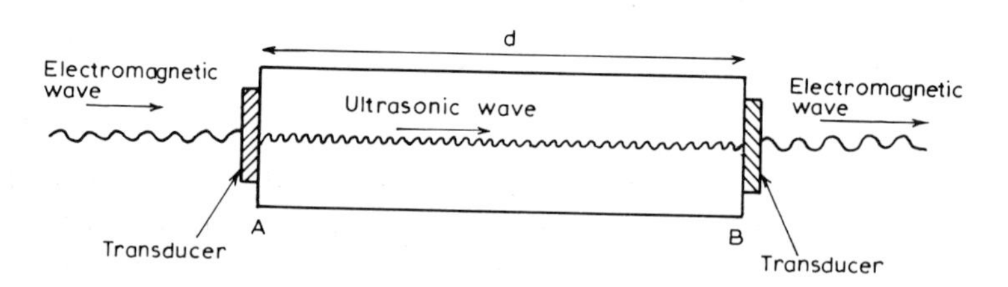

Fig. 2.4. Schematic delay line.

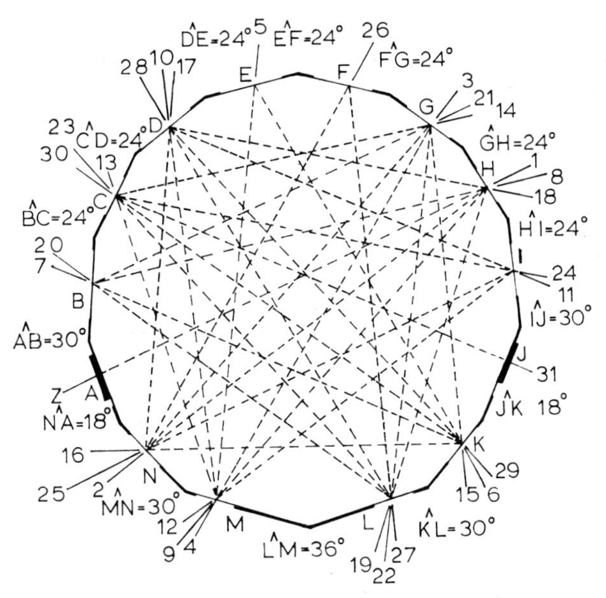

Fig. 2.5. Polygon multiple-reflection delay line. (From D.L. Arenburg, 1954, *I.R.E; Convention Record*, 2, part 6, 63.)

$3000\,\mathrm{m\,s^{-1}}$, the path length needed for a 1 ms delay is only $3000 \times 10^{-3} = 3\,\mathrm{m}$. Multiple reflections, instead of a single transit, enable the length of the line needed for a given delay to be reduced still further (fig. 2.5).

The value of c in eqn. (2.22) appears to be independent of the frequency ν; to a good first approximation this is also true in practice (see Section 7.1). Another way of looking at this would be to say that the speed, and therefore the refractive index, of the ultrasound is independent of the wavelength. Thus, if a beam of ultrasound contains a mixture of frequencies, all components will travel through the medium at the same speed. In practice c is not quite independent of frequency; for example, for a gas the high-frequency limit of c may be typically between 5 and 10% higher than the low-frequency value. The variation of c with frequency is called *dispersion*. If we rearrange eqn. (2.20) we obtain

$$\nu = c/\lambda \tag{2.28}$$

or

$$\omega = ck. \tag{2.29}$$

A graph of ω $(= 2\pi\nu)$ against k $(= 2\pi/\lambda)$ is called a *dispersion relation*. If c is constant, we have a special case and the dispersion relation becomes a straight line (see fig. 2.6). Although the quantities on the right-hand side of eqn. (2.22) are independent of frequency, they are not usually

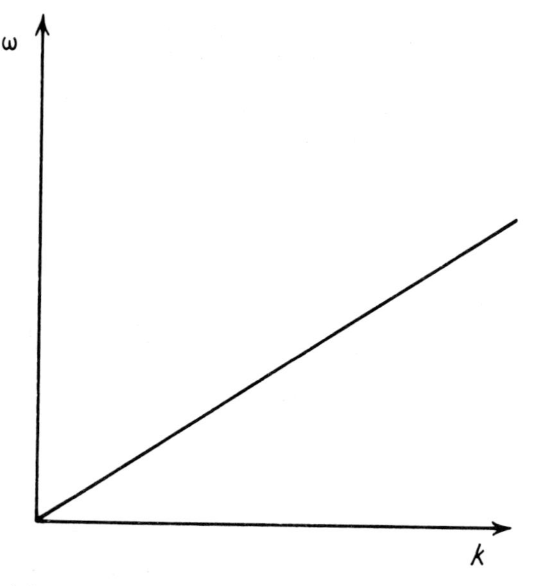

Fig. 2.6. Straight line dispersion relation, ω as a function of k, for an isotropic medium.

independent of temperature. Consequently the value of c will also depend on the temperature.

For a gas we have

$$c = \sqrt{\left(\frac{\gamma p}{\rho}\right)} = \sqrt{\left(\frac{\gamma p V}{M}\right)} = \sqrt{\left(\frac{\gamma R T}{M}\right)} \qquad (2.30)$$

where M is the mass of one mole (in kg); thus, ideally, c is proportional to $\sqrt{T}$ and independent or pressure. For a liquid c is also a function of temperature, see Table 2.2.

2.2. *The Doppler effect*

First, let us consider the propagation in a medium which is at rest. Suppose that a source of ultrasound with frequency ν is moving towards a stationary observer at a velocity v_s. The source is initially at a point S_1 and the observer is at O, where $S_1O = ct$ (see fig. 2.7). Then in time t the source will move a distance $v_s t$ to the point S_2 and the ultrasonic waves that would normally, from a stationary source, occupy the distance S_1O ($= ct$) will be squeezed into the distance S_2O. Thus the wavelength of the ultrasound will be reduced from λ to λ' where

Table 2.2 Speed of sound in sea water as a function of temperature.

Temperature °C	c m s^{-1}
0	1449·5
5	1471·1
10	1490·2
15	1507·1
20	1521·9
25	1534·7
30	1545·9

(Values extracted from G.W.C. Kaye and T.H. Laby, *Tables of Physical and Chemical Constants*, 14th ed. (Longmans, 1973).)

$$\frac{\lambda'}{\lambda} = \frac{S_2O}{S_1O} = \frac{ct - v_s t}{ct} = \frac{c - v_s}{c}, \tag{2.31}$$

likewise the observed frequency ν' is given by

$$\nu' = \frac{c}{c - v_s}\,\nu. \tag{2.32}$$

That is, the frequency of an approaching source appears to be higher than that of the same source at rest. If the source is moving away from the observer, this can be accounted for by changing the sign in the denominator of the right-hand side of eqn. (2.32). We can avoid using two separate formulae by using eqn. (2.32) and adopting a sign convention (see fig. 2.8). Also one can show that if an observer is moving at velocity v_o away from a stationary source, the observed frequency will be

$$\nu' = \frac{c - v_o}{c}\,\nu, \tag{2.33}$$

which is lower than ν. Using the sign convention that v_o is positive if it is in the same direction as c and is negative if it is in the opposite direction (see fig. 2.9) we have, if both source and observer are moving,

$$\nu' = \frac{c - v_o}{c - v_s}\,\nu. \tag{2.34}$$

We can compare this with the case of a moving medium for which the frequency was unaltered and the velocity and wavelength were changed.

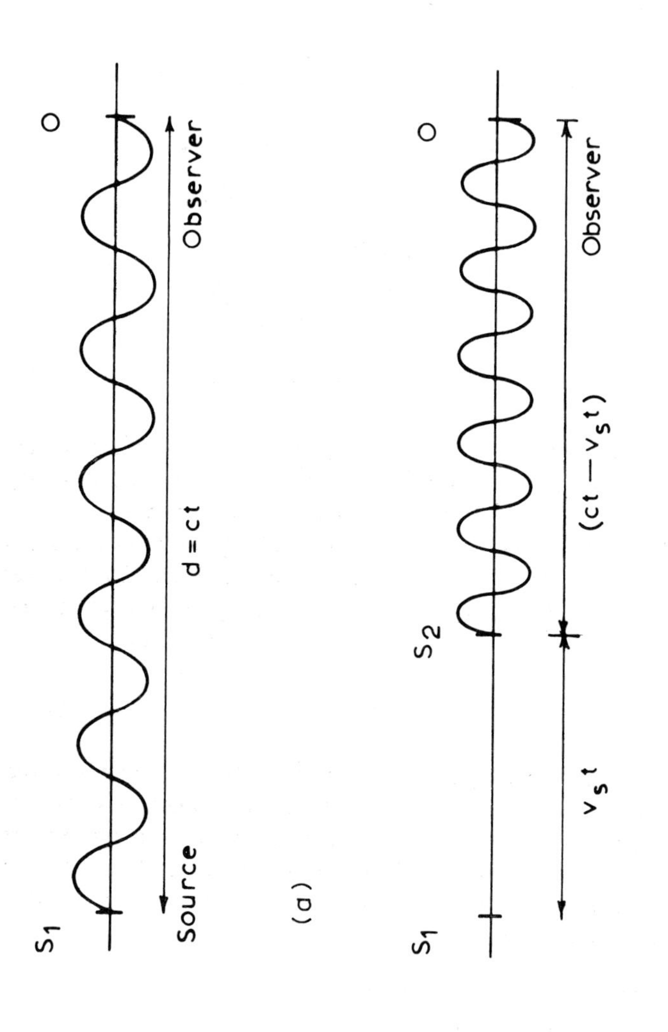

Fig. 2.7 Diagram to illustrate Doppler effect, reduction of wavelength by moving source.

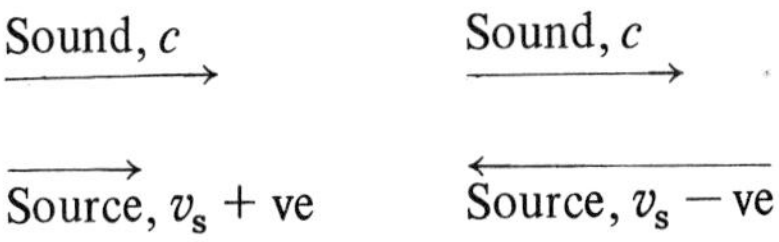

Sound, c $\longrightarrow$ Sound, c $\longrightarrow$

$\overrightarrow{\text{Source}}$, v_s + ve $\overleftarrow{\text{Source}}$, v_s − ve

Fig. 2.8 Sign convention for Doppler effect formula, moving source.

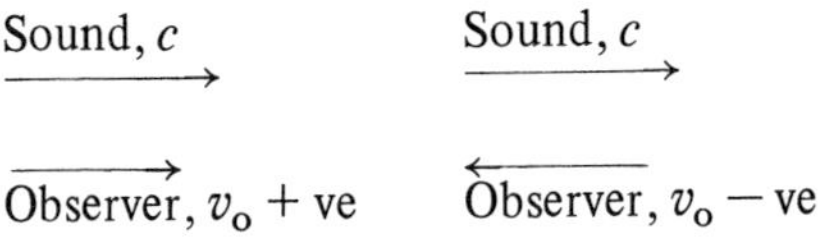

Sound, c $\longrightarrow$ Sound, c $\longrightarrow$

$\overrightarrow{\text{Observer}}$, v_o + ve $\overleftarrow{\text{Observer}}$, v_o − ve

Fig. 2.9. Sign convention for Doppler effect formula, moving observer.

Here, for a moving source or moving observer, it is the velocity which is unaltered and the frequency and wavelength that change.

The change in frequency for the case of eqn. (2.33) is

$$\Delta\nu = \nu' - \nu = \left(\frac{c - v_o}{c}\right)\nu - \nu = -\frac{v_o}{c}\nu. \tag{2.35}$$

A similar formula can be obtained from eqn. (2.34). In all these simplified cases, c, v_s, and v_o are assumed to be the velocities along the line joining the source and the observer (or the components along this line).

In the case of a stationary observer receiving ultrasound from a stationary source, after the ultrasound has been reflected by a mirror moving with velocity v_r (see fig. 2.10), one can regard the image of the source as receding with a velocity $2v_r$, so that eqn. (2.35) gives

$$\Delta\nu = -\frac{2v_r}{c}\nu. \tag{2.36}$$

Thus the ultrasonic Doppler effect provides a direct method of determining the velocity of a moving object. Ultrasonic Doppler measurements can be used, for example, in medical examinations to study movements within the body; they have been used to measure blood-flow rates, to study heart movements, particularly the timing of the heart valves, and in the detection of foetal heart movements.

2.3. *Reflection and refraction of ultrasound*

In optics the 'ray' laws of reflection and of refraction are:

For *reflection* at a plane boundary: (i) the incident ray, the reflected ray, and the normal at the point of incidence all lie in one plane; and (ii) the angle of incidence, i, is equal to the angle of reflection, r, that is

$$i = r. \tag{2.37}$$

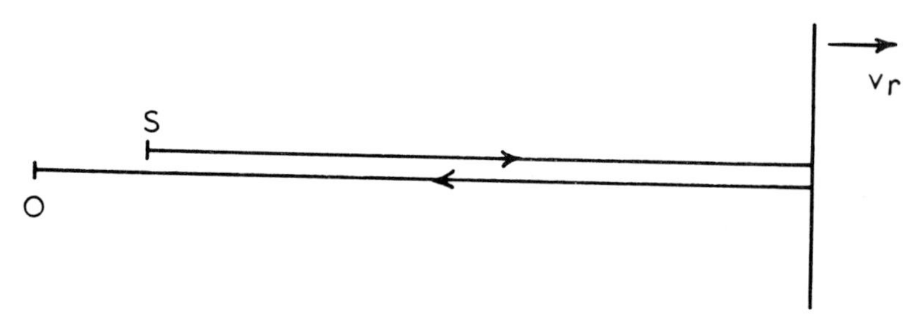

Fig. 2.10 Doppler effect, moving reflector.

For *refraction* at a plane interface between two media (1 and 2): (i) the incident ray, the refracted ray, and the normal all lie in one plane; and (ii) (Snell's Law)

$$n_1 \sin \theta_1 = n_2 \sin \theta_2 \tag{2.38}$$

where θ_1 is the angle of incidence, θ_2 is the angle of refraction, and n_1 and n_2 are called the refractive indices of the two media. According to wave theory $n_1 = c/c_1$ and $n_2 = c/c_2$, where c_1 and c_2 are the speeds of light in the two media and c is the speed of light in a vacuum. *Ray* means the line normal to the wavefronts and so for sound or ultrasound one could replace the word 'ray' by the phrase 'direction of propagation' or 'wave vector **k**'.

The validity of the laws of reflection and refraction has long been established for audible sound, see for example Lord Rayleigh's famous book *The Theory of Sound*, which was first published in 1877.* Lord Rayleigh mentioned a number of basic experiments that had already been performed to investigate the reflection of (audible) sound as well as the refraction of sound by a 'lens' consisting of a balloon filled with CO_2. He also described several experiments that indicate the wave-like nature of ultrasound, using mechanical generators in the form of bird calls, or whistles, producing frequencies of up to 30 kHz or even 50 kHz and using sensitive flames as detectors (see Section 4.1). Apart from direct demonstrations of the laws of reflection and refraction, which Rayleigh took rather for granted for ultrasound, he described some more elaborate experiments.

> By means of a bird-call giving waves of about 1 cm wavelength and a high pressure sensitive flame it is possible to imitate many interesting optical experiments. With this apparatus the shadow of an obstacle so small as the hand may be made apparent at a distance of several feet.
>
> An experiment showing the antagonism between the parts of a wave

* J.W. Strutt, Baron Rayleigh, *The Theory of Sound*, 2nd ed. (Macmillan, 1894, reprinted by Dover Books, 1945).

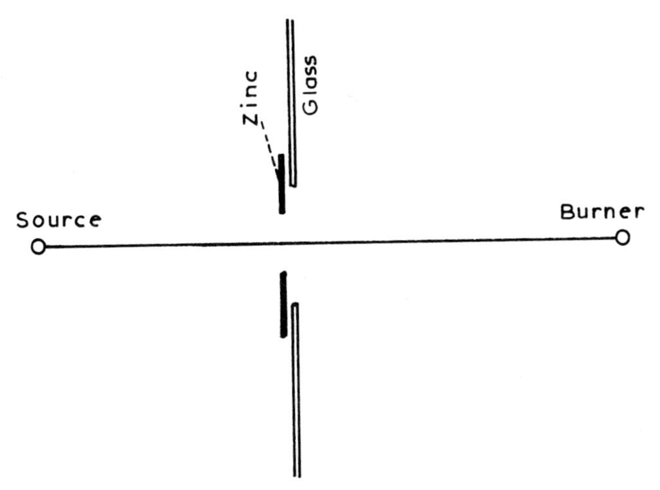

Fig. 2.11. Lord Rayleigh's arrangement for Fresnel's zones experiment using ultrasound. (From J.W. Strutt, Baron Rayleigh, *Theory of Sound*, (Macmillan, London, 1877), reprinted by Dover Books, 1945.)

corresponding to the first and second Fresnel's zones is very effective. A large glass screen [fig. 2.11] is perforated with a circular hole 20 cm in diameter, and is so situated between the source of sound and the burner that the aperture corresponds to the first two zones. By means of a zinc plate, held close to the glass, the aperture may be reduced to 14 cm, and then admits only the first zone. If the adjustments are well made, the flame, unaffected by the waves which penetrate the larger aperture, flares violently* when the aperture is further restricted by the zinc plate. Or, as an alternative, the perforated plate may be replaced by a disc of 14 cm diameter, which allows the second zone to be operative while the first is blocked off. . . .

The process of augmenting the total effect by blocking out the alternate zones may be carried much further. Thus when a suitable circular grating,† cut out of a sheet of zinc, is interposed between the source of sound and the flame, the effect is many times greater than when the screen is removed altogether. As in Soret's corresponding optical experiment, the grating plays the part of a condensing‡ lens.

The focal length of the lens . . . may be written in the form . . .

$$f = \rho^2/n\lambda§.$$

In an actual grating constructed upon this plan eight zones – the first, third, fifth, etc. – are occupied by metal. The radius of the first zone,

* Indicating a greater ultrasonic intensity at the detector (see Section 4.1).
† i.e. a zone plate.
‡ i.e. converging.
§ ρ = radius of zone and n = number of zone.

or central circle, is 7·6 cm, so that $\rho^2/n = 58$. Thus, if $\lambda = 1{\cdot}2$ cm, $f = 48$ cm . . .

The condition of things at the centre of the shadow of a circular disc is still more easily investigated. If we construct in imagination a system of zones beginning with the circular edge of the disc, we see, . . . that the total effect at a point upon the axis, being represented by the half of that of the first zone, is the same as if no obstacle at all were interposed. This analogue of a famous optical phenomenon is readily exhibited. In one experiment a glass disc 38 cm in diameter was employed, and its distances from the source and from the flame were respectively 70 cm and 25 cm.

Nowadays the various experiments to demonstrate the laws of reflection and refraction of ultrasound and to demonstrate its other wave-like properties would be modified by replacing mechanical generators and sensitive-flame detectors by piezoelectric or magnetostrictive transducers and the associated electronics (see Chapter 4).

We now turn to the consideration of the determination of reflection and transmission coefficients. Consider longitudinal waves travelling in a medium A and incident normally on a plane interface between A and a second medium B. Suppose that the displacements $u_i(x, t)$, $u_r(x, t)$ and $u_t(x, t)$ in the incident, reflected, and transmitted waves are written as

$$\left.\begin{aligned} u_i(x, t) &= a_i \exp[i(\omega t - k_A x)] \\ u_r(x, t) &= a_r \exp[i(\omega t + k_A x)] \\ u_t(x, t) &= a_t \exp[i(\omega t - k_B x)] \end{aligned}\right\} \qquad (2.39)$$

where $k_A = 2\pi/\lambda_A$ and $k_B = 2\pi/\lambda_B$ and λ_A and λ_B are the wavelengths in A and B. The intensities a_r^2 and a_t^2, of the reflected and transmitted waves, can be determined by using the boundary conditions that apply at the interface (see fig. 2.12). These boundary conditions are:

(*a*) The displacement at the boundary is the same in both media, in other words the two media stay in contact, i.e.

$$u_i(x, t) + u_r(x, t) = u_t(x, t) \qquad (2.40)$$

at $x = 0$ and for all t.

(*b*) The stress at the boundary is the same in both media, i.e.

$$\mathscr{E}_A \frac{\partial u_i(x, t)}{\partial x} + \mathscr{E}_A \frac{\partial u_r(x, t)}{\partial x} = \mathscr{E}_B \frac{\partial u_t(x, t)}{\partial x} \qquad (2.41)$$

at $x = 0$ and for all t, where $\mathscr{E}_A$ and $\mathscr{E}_B$ are the appropriate moduli of elasticity for A and B. By substituting from equation (2.39) into equations (2.40) and (2.41) and solving the resulting equations to give a_r^2 and a_t^2 in terms of a_i we obtain

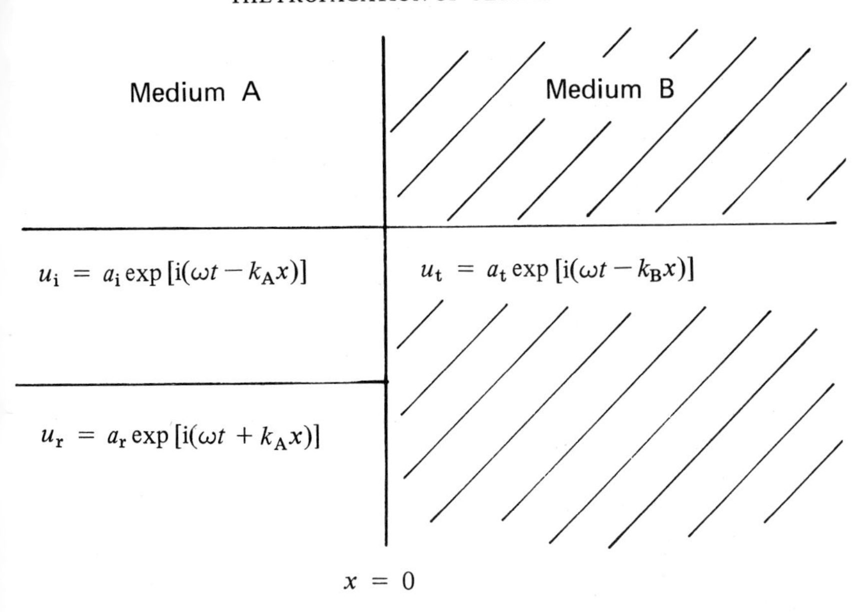

Fig. 2.12. Notation for reflection and transmission of waves at a boundary.

$$\left.\begin{aligned} a_r^2 &= \left(\frac{\mathscr{E}_A k_A - \mathscr{E}_B k_B}{\mathscr{E}_A k_A + \mathscr{E}_B k_B}\right)^2 a_i^2 \\ a_t^2 &= \left(\frac{2\mathscr{E}_A k_A}{\mathscr{E}_A k_A + \mathscr{E}_B k_B}\right)^2 a_i^2 \end{aligned}\right\}. \tag{2.42}$$

We can obtain convenient alternative expressions for a_r^2 and a_t^2 by introducing the idea of 'acoustic impedance' and replacing $\mathscr{E}_A k_A$ and $\mathscr{E}_B k_B$. Using $c = \omega/k$ from equation (2.20) and then using equation (2.22), which gives $\mathscr{E} = c^2\rho$, we can write

$$\mathscr{E}k = \mathscr{E}\omega/c = c^2\rho\omega/c = \rho c\omega. \tag{2.43}$$

The quantity ρc is taken to be the *characteristic acoustic impedance* w of a medium.

The physical significance of w will begin to become apparent if we substitute the characteristic acoustic impedance $w_A = \mathscr{E}_A k_A$ and $w_B = \mathscr{E}_B k_B$ for the two media A and B, respectively, into equation (2.42); this gives

$$\left.\begin{aligned} a_r^2 &= \left(\frac{w_A - w_B}{w_A + w_B}\right)^2 a_i^2 \\ a_t^2 &= \left(\frac{2w_A}{w_A + w_B}\right)^2 a_i^2 \end{aligned}\right\}. \tag{2.44}$$

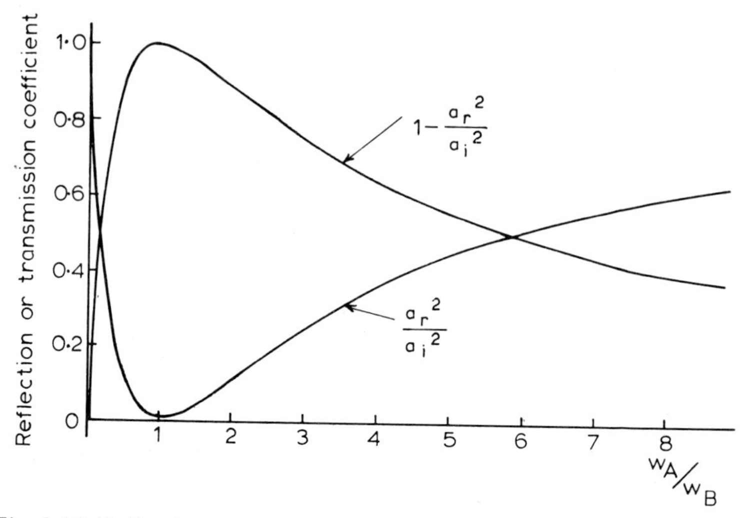

Fig. 2.13 Reflection and transmission coefficients as a function of w_A/w_B.

From this equation we see that if $w_B \gg w_A$ then

$$\left.\begin{aligned} a_r^2 &\sim a_i^2 \\ a_t^2 &\sim 0 \end{aligned}\right\}. \tag{2.45}$$

This means that if an ultrasonic wave is travelling in a medium of low acoustic impedance and encounters a boundary with a second medium of high acoustic impedance the energy is almost entirely reflected; this is in agreement with the idea of impedance used in other contexts, particularly in connection with electromagnetic waves. Alternatively, if w_A and w_B are very nearly equal, equation (2.44) means that

$$\left.\begin{aligned} a_r^2 &\sim 0 \\ a_t^2 &\sim a_i^2 \end{aligned}\right\} \tag{2.46}$$

so that almost all of the energy is transmitted and very little is reflected. If $w_B \ll w_A$

$$a_r^2 \sim a_i^2 \tag{2.47}$$

and, again, nearly all the energy is reflected (see fig. 2.13). It should be mentioned that the energy carried by the incident wave, the reflected wave and the transmitted wave is $\frac{1}{2}\omega^2 w_A a_i^2$, $\frac{1}{2}\omega^2 w_A a_r^2$ and $\frac{1}{2}\omega^2 w_B a_t^2$, respectively (eqn. (1.13)); so

$$\textit{fraction of energy reflected} = \frac{a_r^2}{a_i^2}$$

and (2.48)

$$\textit{fraction of energy transmitted} = \frac{w_B a_t^2}{w_A a_i^2}$$

$$= 1 - \frac{a_r^2}{a_i^2} \tag{2.49}$$

$$\left(\neq \frac{a_t^2}{a_i^2}\right).$$

To obtain the maximum transmission a_r should be as small as possible; this is when $w_A = w_B$ and the two media are then said to be *acoustically matched*. However, in many situations w_A and w_B may have quite different values and, moreover, we may not be free to alter w_A and w_B. In this case the amount of energy that is reflected may be reduced by having a gradual change in the impedance, ρc, at the boundary or by having a suitable intermediate layer. The question of acoustic matching will be discussed again in Section 4.4. Some values of the acoustic impedance for a variety of important materials are listed in Table 2.1.

The stress $(= F(x)/A)$ associated with a longitudinal elastic wave is related to the displacement via eqn. (2.12). We can think of this stress as the excess pressure amplitude $p(x, t)$ associated with the wave so that, using eqn. (2.12),

$$p(x,t) = E\frac{\partial u(x,t)}{\partial x}. \tag{2.50}$$

For a wave with displacements given by

$$u(x,t) = a\exp[\mathrm{i}(\omega t - kx)], \tag{2.17}$$

we therefore have

$$p(x,t) = Ea(-\mathrm{i}k)\exp[\mathrm{i}(\omega t - kx)] = -\mathrm{i}kEu(x,t). \tag{2.51}$$

Regarding the $-\mathrm{i}$ as a phase factor, this means that the ratio

$$\frac{\text{excess pressure}}{\text{particle displacement}} = Ek \tag{2.52}$$

and as we have indicated before, $Ek = \rho c$ is called the acoustic impedance,

w.* It is eqn. (2.52) rather than $w = \rho c$, which is really the definition of the acoustic impedance. This definition is the analogue, for ultrasound, of the equation that defines the electrical impedance of a dielectric medium for the propagation of electromagnetic waves.

The arguments that we used in finding the reflection and transmission coefficients for normal incidence in eqn. (2.42) can be repeated for oblique incidence, that is for $i \neq 0$. This is done by fitting the boundary conditions on the displacements and on the pressure at the boundary. However, the details are substantially more complicated for oblique incidence than for normal incidence, because 'mode conversion' often occurs at an interface between two media. Thus an obliquely incident wave of one type, either longitudinal or transverse, may generate at the interface both longitudinal and transverse components in both the reflected and refracted waves. Another point that should be mentioned at this stage is that the sense of refraction of an ultrasonic beam crossing a boundary may be the reverse of the sense for a light beam crossing a similar boundary. For example, ultrasound passing from a liquid to a solid generally experiences an increase in velocity (see Table 2.1) and so bends away from the normal; thus for ultrasound a solid converging lens immersed in a liquid will usually have to be concave (see fig. 7.14).

2.4. *Lattice vibrations – phonons*

In the earlier sections of this chapter we have adopted a macroscopic approach to the propagation of sound, or of ultrasound, in a medium and we have treated the medium as a continuum. Thus we have regarded the medium rather like a jelly with its elastic properties specified by the appropriate elastic moduli or, for anisotropic materials, by the appropriate tensor components (see Section 9.1). In this analogy a solid corresponds to a very stiff jelly, while liquids and gases correspond to much less stiff jellies. In this approach the fact that the medium is not really a continuum, but is made up of atoms, or ions or molecules, was completely ignored. The displacement $u(x, t)$ given by eqn. (2.17) or eqn. (2.18) is a continuous function of position, x; this means that a value of $u(x, t)$ can be assigned

* The reason why we did not attempt a quantitative discussion of radiation pressure in Section 1.3 is that, in spite of appearances, eqn. (2.51) does not really represent a linear relationship between $p(x, t)$ and $u(x, t)$. This can be seen if eqn. (2.51) is rewritten as

$$p(x, t) = -\mathrm{i}\omega\rho c \, u(x, t)$$

because the density ρ is not really constant since in regions of compression the density is greater than the mean value, and in regions of rarefaction the density is smaller than the mean value.

for every value of x (see fig. 2.14(a)). Let us take into account the fact that in practice a medium which transmits ultrasound is not continuous, but is made up of atoms. Then it is only really meaningful to speak about the displacement $u(x, t)$ for the atoms and not for the empty spaces in between the atoms. Thus the smooth curve of $u(x, t)$ in fig, 2.14(a) should be replaced by a sequence of dots (see fig. 2.14(b)). However, if the interatomic spacing a is very much smaller than the wavelength λ of the ultrasound it will be a good approximation to regard the sequence of dots as a smooth curve (see fig. 2.14(a)). This means that provided

$$\lambda \gg a \tag{2.53}$$

the approach which was adopted in the earlier sections of this chapter, and which was based on a continuum model, will be satisfactory. However, if λ is reduced, relative to a, the replacement of the sequence of dots for $u(x, t)$ by a continuous curve becomes much less reasonable (see fig. 2.14(b) and (c)).

Typical values of a, the interatomic spacing, in a solid are of the order of one or two Å, say $1{\cdot}5 \times 10^{-10}$ m. Suppose that we adopt a slightly arbitrary limit and take the criterion in eqn. (2.53) for the validity of the continuum model to mean that λ must not be smaller than $10a$, i.e. $\lambda > 1{\cdot}5 \times 10^{-9}$ m. This corresponds to an upper limit to the frequency of ultrasonic waves for which the continuum model is valid; if we assume a value of $c \sim 1000\ \mathrm{m\,s^{-1}}$ this gives a limiting frequency of

$$\nu = \frac{c}{\lambda} = \frac{1000}{1{\cdot}5 \times 10^{-9}} = 6 \times 10^{11}\ \mathrm{Hz}. \tag{2.54}$$

This frequency is in the microwave acoustic range of frequencies (see fig. 1.2). Therefore for frequencies lower than this limit the elastic continuum, or jelly, model is valid.

In this section we shall consider a microscopic approach to the propagation of ultrasound, and the standing ultrasonic waves, in a material medium. One has to study the propagation of energy through an array of atoms, ions, or molecules forming a gas, liquid, or solid. This energy is transmitted in the form of vibrations of these atoms, ions, or molecules and in this section it is our purpose to explore this microscopic approach to the problem of the transmission of elastic waves. For simplicity we shall restrict our discussion to crystalline solids because, for a quantitative microscopic discussion of elastic vibrations, they are easier to handle than other media.

Before leaving the continuum model, however, we should recall some early theoretical work on the specific heats of solids. If the temperature of a solid is raised, the average kinetic energy of vibration of the molecules in

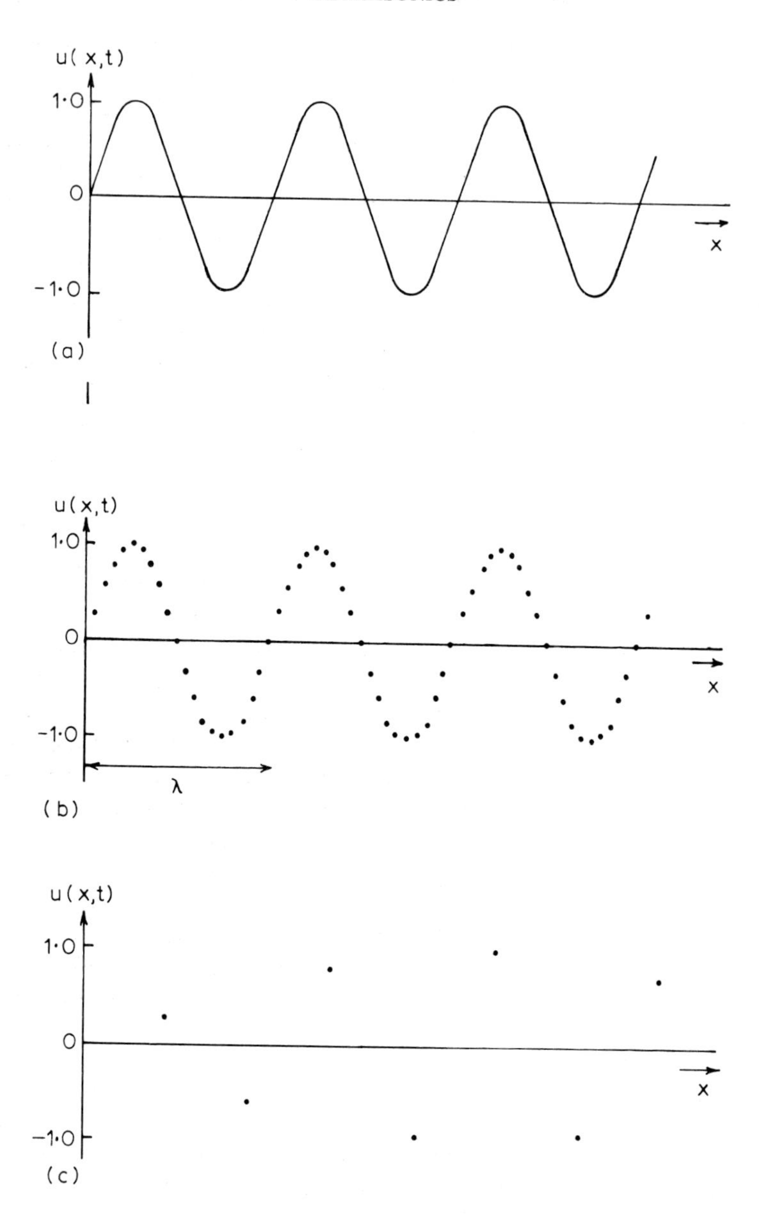

Fig. 2.14 $u(x, t)$ for (*a*) a continuous medium (jelly) and for a (one-dimensional) medium composed of atoms separated by a distance a from each of their nearest neighbours (*b*) $a \ll \lambda$, and (*c*) $a \sim \lambda$.

that solid will be increased. The first theoretical attempts that met with even a modest degree of success in explaining the behaviour of the specific heat of a solid as a function of temperature were made by Einstein in 1907 and by Debye in 1912. Instead of considering the vibrations of the individual atoms in the solid, it is advantageous to consider the normal modes of vibration of the complete array of atoms in the solid. Since these normal modes are standing sinusoidal waves, this corresponds to a three-dimensional Fourier analysis of the vibrations of all the individual atoms. For many years there was comparatively little quantitative knowledge about the forces that exist between the atoms in a solid. Without the knowledge of these forces one cannot obtain the equations of motion and so one cannot calculate the frequencies of vibration of the molecules or the frequencies of the normal modes of a solid specimen. The alternative would be to try to measure ν experimentally, but this only became feasible after the Second World War when inelastic neutron-scattering techniques were introduced. Einstein made the simplest possible assumption, namely that all the normal modes have the same frequency. Debye improved on this by regarding the normal modes as standing sinusoidal waves in a continuous medium, the behaviour of which was completely specified by the moduli of elasticity, or elastic stiffness constants, of the crystal as measured macroscopically. These normal modes are then none other than the waves of sound or ultrasound with which we have been concerned in the earlier sections of this chapter. The only difference is that they are standing waves, instead of being progressive waves. In other words, the random thermal vibrations of the molecules in a solid are regarded as being equivalent to having a large number of standing waves of sound and of ultrasound in the solid. Although it does not give a perfect agreement with experimental results, the Debye theory gives a good approximation to the measured values of the vibrational specific heat of a solid as a function of temperature. The discrepancies between theory and experiment arise from the fact that it is necessary to include in the calculations some ultrasonic waves of sufficiently short wavelength that the criterion in eqn. (2.54) for the validity of the continuum model is violated.

One specially important feature of the continuum model is that the velocity of propagation, given by eqn. (2.22) or eqn. (2.32), will be independent of the frequency of the ultrasound. We have seen in fig. 2.6 that this means that the dispersion relation, that is ν, or ω, plotted as a function of $k(= 2\pi/\lambda)$ is a straight line where

$$\omega = ck. \tag{2.29}$$

These features, namely that c is independent of frequency and that the graph of ν (or ω) against k is a straight line, only partially survive once we

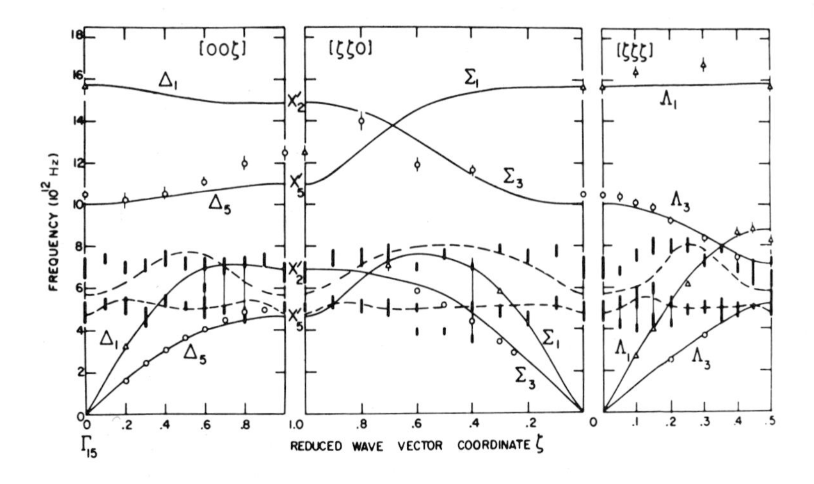

Fig. 2.15. Phonon dispersion relations for some important directions of propagation in CoO at 110 K, (From J. Sakurai, W.J.L. Buyers, R.A. Cowley and G. Dolling, 1968, *Phys. Rev.*, **167**, 510.).

abandon the continuum model. When inelastic neutron-scattering techniques were introduced it became possible to determine the dispersion relations, that is curves of ν (or ω) against k, experimentally, for very short wavelengths (i.e. large k), and to detect departures from the linear relation given in eqn. (2.29) (see fig. 2.15). The branches which pass through Γ ($k = 0$, i.e. very long wavelength) are approximately linear near to Γ; they are sometimes referred to as *acoustic branches* of the dispersion relations, or simply as *acoustic modes*. The limited linear regions of the acoustic branches can indeed be determined experimentally from measurements of the ultrasonic velocity. The other branches of the dispersion relations, which do not pass through Γ, are commonly referred to as *optic branches* of the dispersion relations, or simply as *optic modes*; the origin of this nomenclature lines in the fact that the frequencies involved are, typically, of the order of magnitude of optical frequencies. It is not unreasonable to say that nowadays experimentally-determined dispersion relations for a solid, determined principally from inelastic neutron scattering experiments, are used to study the nature of the forces between the atoms in the solid.

To develop a microscopic theory of elastic vibrations in a solid it is necessary to be able to write down mathematical expressions for the forces between the atoms and then one can use Newton's second law of motion to obtain the equations of motion for the vibrating atoms. For a

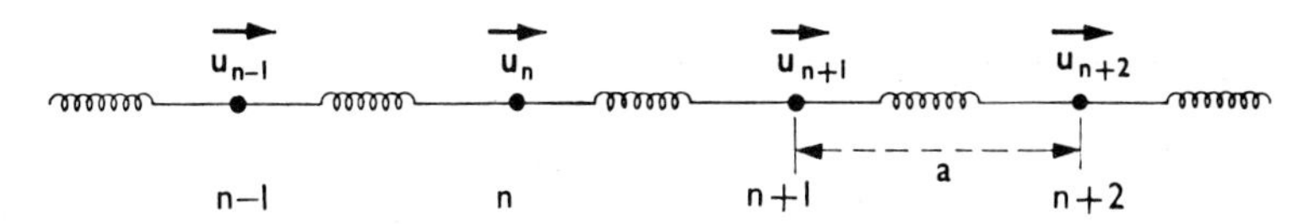

Fig. 2.16 One-dimensional model of vibrations of atoms in a solid.

three-dimensional system the algebra becomes rather complicated and so we shall just illustrate what is involved with a simple one-dimensional model, which is illustrated in fig. 2.16. This consists of a chain of identical particles, each of mass m, joined by identical springs, each with force constant β, so that the equilibrium positions of the particles are equally spaced, a apart. Suppose that $u_{n-1}(t)$, $u_n(t)$, and $u_{n+1}(t)$ are the displacements of the particles labelled by $n-1$, n, and $n+1$, respectively. Then the extension of the spring joining particles n and $n+1$ is $u_{n+1}(t)-u_n(t)$ and the additional tension in the spring will be equal to $\beta\{u_{n+1}(t)-u_n(t)\}$. Similarly, the additional tension in the spring between particles $n-1$ and n is $\beta\{u_n(t)-u_{n-1}(t)\}$. Therefore we can use Newton's second law of motion for the nth particle

$$\beta\{u_{n+1}(t)-u_n(t)\}-\beta\{u_n(t)-u_{n-1}(t)\} = m\,\ddot{u}_n(t) \tag{2.55}$$

or

$$m\,\ddot{u}_n(t) = \beta\{u_{n+1}(t)-2u_n(t)+u_{n-1}(t)\}. \tag{2.56}$$

We take a trial solution of the form

$$u_n(t) = A\exp[\mathrm{i}(\omega t+kna)] \tag{2.57}$$

where na corresponds most closely to the continuous variable x used in Section 2.1; substituting this into eqn. (2.56) gives

$$-\omega^2 m = \beta[\exp(\mathrm{i}ka)-2+\exp(-\mathrm{i}ka)] = \beta[\exp(\tfrac{1}{2}\mathrm{i}ka)-\exp(-\tfrac{1}{2}\mathrm{i}ka)]^2$$
$$= -4\beta\sin^2\tfrac{1}{2}ka. \tag{2.58}$$

Therefore wave-like solutions of the form given in eqn. (2.57) are possible provided

$$\omega = \pm 2\sqrt{\left(\frac{\beta}{m}\right)}\sin\tfrac{1}{2}ka. \tag{2.59}$$

This equation gives the dispersion relation for wave-like sets of displacements travelling along the one-dimensional system illustrated in fig. 2.17.

The graph of ω as a function of k given by eqn. (2.59) is shown in fig. 2.17. Although this is only a crude model of a real solid, the dispersion relation in fig. 2.17 does exhibit some important features that are also present in the dispersion relations for real three-dimensional crystals. First, for very small k, that is for $\lambda \gg a$, the dispersion curve is approximately a

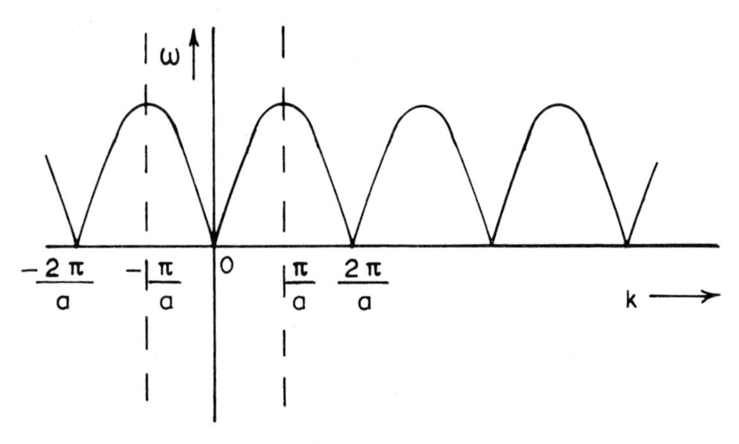

Fig. 2.17. ω as a function of k for the one-dimensional structure illustrated in fig. 2.16.

straight line ($\sin \frac{1}{2}ka \sim \frac{1}{2}ka$) which, of course, is consistent with what we would obtain from eqn. (2.29) based on the continuum model. Secondly, ω is a periodic function of k; this arises because of the regular periodic nature of the equilibrium positions of the masses in the original system. This means that it is only necessary to consider wave vectors k within the range $-\pi/a \leqslant |\mathbf{k}| \leqslant \pi/a$ because any k outside this range is related by an amount $2\pi N/a$ ($N =$ integer) to some k which has the same ω and which is within this range. If the equilibrium positions of the masses did not form a regular periodic array, ω would not be a periodic function of k. This would correspond, in three dimensions, to a non-crystalline solid.

The macroscopic approach, based on the idea of a continuum, which was used in the earlier sections of this chapter has principally been used in connection with propagating waves, while the microscopic approach that we have just outlined in this section has principally been used in connection with standing waves. It should, perhaps, be emphasized that this difference is not intrinsic to the two approaches. In each case one derives the appropriate equation of wave motion and then determines the solutions of this equation. Then in each case one can choose either the stationary-wave or propagating-wave solutions. Which kind of solutions are chosen depends on physical considerations. The propagating-wave solutions correspond to situations in which sound or ultrasound is produced by a generator and transmitted through a medium. Generally the wavelength is rather long so that

$$\lambda \gg a \tag{2.53}$$

and the macroscopic approach is relevant, although at the very high ultrasonic frequencies (microwave acoustic frequencies) which are now used in some scientific work, the values of λ may no longer satisfy (2.53). The standing-wave solutions which apply to the random thermal vibrations of a solid may be of long wavelength ($\lambda \gg a$) or they may be of shorter wavelength ($\lambda \sim a$). At low temperatures it is principally the long-wavelength modes which are excited and so at low temperatures the macroscopic, or continuum, approach may be used. But at higher temperatures there will also be a considerable amount of excitation of the short-wavelength modes so that it then becomes necessary to use the microscopic approach.

One final comment should perhaps be made. The energy in a wave of sound, or of ultrasound, is quantized in much the same way that the energy in a light wave is quantized. Just as the energy in a photon (a quantum of light) of frequency ν is given by

$$E = h\nu \tag{2.60}$$

so also the energy in a phonon (a quantum of vibrational energy) is given by the same formula, or in terms of ω by

$$E = h\frac{\omega}{2\pi} = \hbar\omega. \tag{2.61}$$

For low frequencies, that is for audible sound and for the lower ultrasonic frequencies, the energy of a quantum is so very small that any quantum effects are negligible. It is only at extremely high ultrasonic frequencies, that is microwave acoustic frequencies, that quantum effects become important (see Section 9.3).

3. The attenuation of ultrasound

3.1. *Attenuation coefficients*

Suppose that a parallel beam of sound or ultrasound is propagating in the x direction in some medium and that the intensity, or the excess-pressure amplitude, is measured at x_1 and x_2 (fig. 3.1). Then the word 'attenuation' is used to describe the total reduction in the intensity, or excess-pressure amplitude, of the beam between x_1 and x_2. This attenuation will result from

(1) the absorption of energy by the medium between x_1 and x_2, and

(2) the deflection of energy from the path of the beam by reflection, refraction, diffraction, and scattering.

Absorption involves the conversion of sound or ultrasound into some other form of energy, whereas in the case of reflection, refraction, diffraction, and scattering the sound or ultrasound is simply travelling in some other direction. The absorption will depend on the nature of the medium between x_1 and x_2 and can, therefore, furnish information about the physical properties of the medium. The reflection, refraction, diffraction, and scattering losses depend on the geometry of the system as well as on the physical properties of the medium or media. Reflection and refraction will occur at boundaries between regions with different acoustic impedances (see Section 2.3). Diffraction will occur at barriers that are interposed in the path of the beam. Scattering losses are characteristic of the structure of the material; for example, in a polycrystal they will depend on the grain size.

For a beam of sound or ultrasound, the diminution $-\delta I$ in the intensity I on travelling a distance δx in the medium is proportional to I and to δx. Thus

$$-\delta I = \alpha_I I \delta x \qquad (3.1)$$

and so

$$\frac{\mathrm{d}I}{\mathrm{d}x} = -\alpha_I I \qquad (3.2)$$

where α_I is a constant. Thus

$$I = I_0 \exp(-\alpha_I x). \qquad (3.3)$$

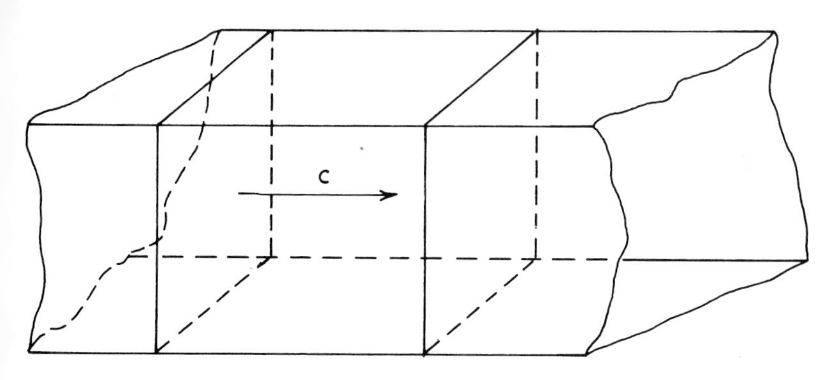

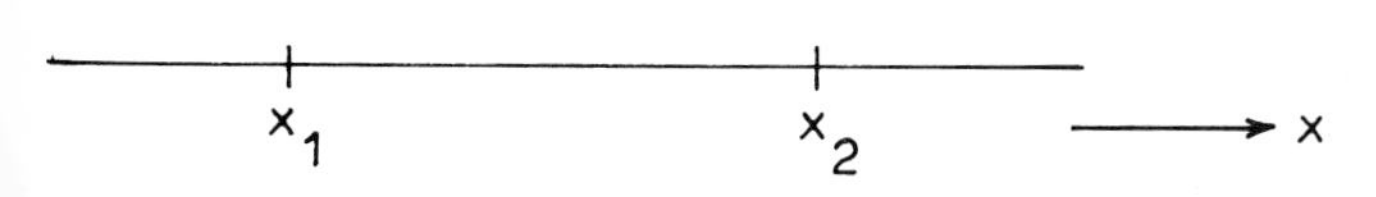

Fig. 3.1.

Alternatively, one can consider the excess pressure p such that

$$p = p_0 \exp(-\alpha_p x). \tag{3.4}$$

α_p is called the attenuation coefficient of the medium.

The excess pressure $p(x, t)$ is proportional to the displacement $u(x, t)$ (see eqn. (2.51)). Since the intensity of an ultrasonic wave is proportional to the square of the displacement, this means that $I \propto \rho^2$. From eqns. (3.3) and (3.4) we have

$$\exp(-\alpha_I x) = \frac{I}{I_0} = \frac{p^2}{p_0^2} = \exp(-2\alpha_p x)$$

and therefore

$$\alpha_I = 2\alpha_p. \tag{3.6}$$

In Chapter 2 we assumed that the wave vector k of an ultrasonic wave was real rather than complex. However, if we write it as a complex quantity

$$k = \kappa - \mathrm{i}\alpha_p \tag{3.7}$$

and substitute into eqn. (2.51), we obtain

$$\begin{aligned} p(x, t) &= p \exp\{\mathrm{i}[\omega t - (\kappa - \mathrm{i}\alpha_p)x]\} \\ &= \exp(-\alpha_p x)\{p \exp[\mathrm{i}(\omega t - \kappa x)]\} \end{aligned} \tag{3.8}$$

so that the pressure amplitudes at x_2 and x_1 will satisfy eqn. (3.4). This means that we can regard α_p as the imaginary component of a complex wave vector $\mathbf{k}$ of an ultrasonic wave.

Attenuation is measured in terms of nepers (Np) or of decibels (dB) which are defined as follows:

The 'power level' is defined as

$$\text{power level} = \log_e \left(\frac{p_0}{p}\right) \text{Np} = 10 \log_{10} \left(\frac{I_0}{I}\right) \text{dB}. \qquad (3.9)$$

But $I/I_0 = p^2/p_0^2$ and $\log_{10} x \doteqdot 0{\cdot}4343 \log_e x$ so that

$$10 \log_{10} \left(\frac{I_0}{I}\right) \text{dB} = 20 \log_{10} \left(\frac{p_0}{p}\right) \text{dB} = 20 \times 0{\cdot}4343 \log_e \left(\frac{p_0}{p}\right) \text{dB}$$

$$= 8{\cdot}686 \log_e \left(\frac{p_0}{p}\right) \text{dB}. \qquad (3.10)$$

Therefore

$$1 \text{ Np} = 8{\cdot}686 \text{ dB}. \qquad (3.11)$$

If we take natural logarithms of both sides of eqn. (3.4) we obtain

$$\alpha_p x = \log_e \left(\frac{p_0}{p}\right) \text{Np} \qquad (3.12)$$

or, alternatively, if we take logarithms to base 10

$$\alpha_p x = 20 \log_{10} \left(\frac{p_0}{p}\right) \text{dB}. \qquad (3.13)$$

Similarly, using equation (3.3) we obtain

$$\alpha_I x = \log_e \left(\frac{I_0}{I}\right) \text{Np} \qquad (3.14)$$

or

$$\alpha_I x = 20 \log_{10} \left(\frac{I_0}{I}\right) \text{dB}. \qquad (3.15)$$

Since $\alpha_p x$ and $\alpha_I x$ are pure numbers, the units of α_p, or of α_I, are therefore Np m^{-1} (etc.) or dB m^{-1} (etc.). Historically the neper is the older unit, but both are now used extensively. If one is only concerned with approximate values then it is convenient to modify eqn. (3.11) to give

$$1 \text{ Np cm}^{-1} = 0{\cdot}8686 \text{ dB mm}^{-1} \qquad (3.16)$$

so that the numerical value of an attenuation coefficient will be of the same order of magnitude whether it is expressed in Np cm^{-1} or in dB mm^{-1}.

Some measured values of attenuation coefficients are given in Table 3.1. For any given material α_I, or α_p, can generally be expected to be a function

Table 3.1. Measured attenuation coefficients α_p.

(a) Gases (at 0°C)	α_p/ν^2 10^{-11} Np m^{-1} s^2	
Air (dry)	1.85	
(b) Liquids (at 20°C)	α_p/ν^2 10^{-15} Np m^{-1} s^2	
Distilled water	25	
Acetone	54	
Carbon tetrachloride	540	
Ethanol	52	
Glycerol	2000	
Mercury	5.5	
(c) Solids (at ~ 20°C) (for longitudinal bulk waves)	Frequency MHz	α_p Np m^{-1}
Aluminium	10	0.40
Crown glass	10	2
Perspex	2.5	57

(Values taken from G.W.C. Kaye and T.H. Laby, *Tables of Physical and Chemical Constants*, 14th edition (Longmans, 1973).)

of the frequency; thus we should write the attenuation coefficient as $\alpha_I(\nu)$, or $\alpha_p(\nu)$. For technical reasons, which should become apparent in the next two chapters, it is not very easy to make accurate measurements of α_I, or of α_p, for a medium. On the other hand, one can quite easily perform measurements of relative attenuation, that is variations in α_I, or in α_p, as one external parameter, such as the temperature or an external magnetic field or the frequency of the ultrasound, is varied; in several scientific applications this is very useful (see Chapter 9).

Several physical processes which may contribute to the absorption of ultrasound as it passes through a medium, and the ones that are actually important may be different for different types of media. These processes are fairly well understood in the case of gases and of liquids and quite good quantitative agreement between theory and experiment can be obtained. For solids it is known qualitatively what mechanisms are involved in the absorption of sound or of ultrasound, but it is very difficult to make accurate quantitative predictions of the various contributions to the absorption coefficient that arise from these mechanisms.

Table 3.2. Maximum noise levels recommended by the Environmental Protection Agency for the protection of public health and welfare.

Effect	Level	Area
Hearing loss	$L_{eq(8)} \leqslant 75$ dB	Occupational and educational settings
	$L_{eq(24)} \leqslant 70$ dB	All other areas
Outdoor activity	$L_{dn} \leqslant 55$ dB	Outdoors in residential areas and farms and other outdoor areas where people spend widely varying amounts of time and other places in which quiet is a basis for use
	$L_{eq(8)} \leqslant 55$ dB	Outdoor areas where people spend limited amounts of time such as school yards, playgrounds, etc.
Indoor activity interference and annoyance	$L_{dn} \leqslant 45$ dB	Indoor residential areas
	$L_{eq(24)} \leqslant 45$ dB	Other indoor areas with human activities such as schools, etc.

Notes.
$L_{eq(8)}$ and $L_{eq(24)}$ are the equivalent noise levels averaged over 8 hours and 24 hours, respectively. The day–night sound level L_{dn} is a weighted mean of day-time and night-time levels.

We have introduced the decibel and the neper with reference to the attenuation of sound, or of ultrasound, as it travels through a medium. However, outside the context of strictly scientific attenuation measurements, people are more likely to be aware of these units, particularly the decibel, in connection with various 'environmental' measurements of sound levels. For example, sound levels measured in the vicinity of an urban motorway or near a busy airport are likely to be expressed in decibels. Such measurements are, essentially, relative measurements. By analogy with eqn. (3.9) we can give a quantitative measure of the relative intensities, I_1 and I_2, of two sounds by saying that

$$\text{difference of intensity} = 10 \log_{10}\left(\frac{I_1}{I_2}\right) \text{dB}. \tag{3.17}$$

In order to use what is, essentially, a relative scale to specify absolute values of sound intensities it is necessary to have some 'reference intensity'. The reference that is chosen is an intensity of 10^{-12} W m^{-2}, which is close

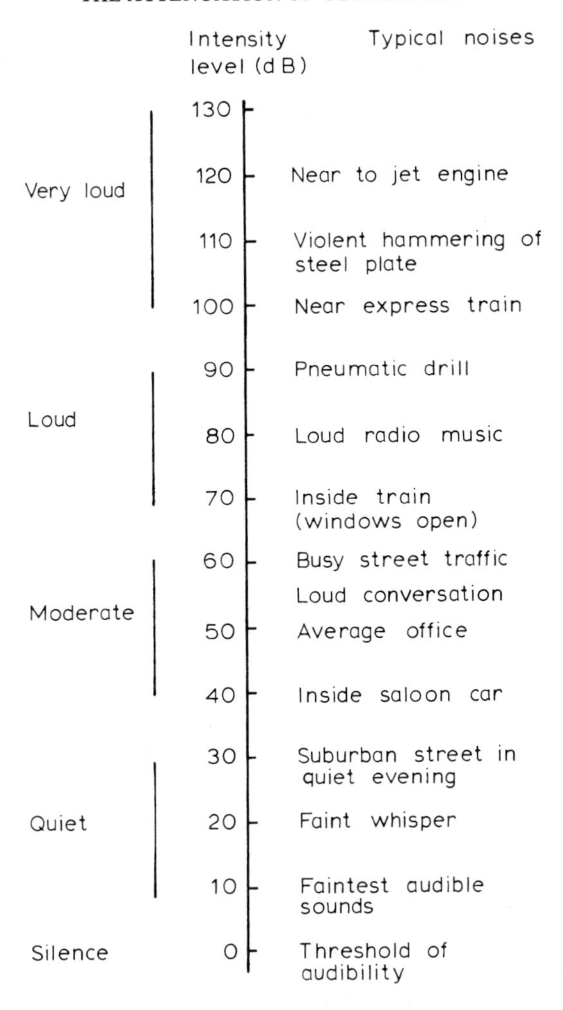

Fig. 3.2. Approximate intensity levels of some typical noises.

to the threshold of audibility for an average human ear at a frequency of 1 kHz. The intensity levels, on this basis, of various well-known sounds are given in fig. 3.2. For audible sounds a change of 1 dB is about the smallest change that can be detected by the human ear, while the complete range of intensity to which the ear can respond is ~ 120 or 130 dB. Values of permitted maximum noise levels recommended by the Environmental Protection Agency are given in Table 3.2.

It might have seemed more natural to express the intensity I of sound or of ultrasound directly in terms of energy units, rather than in terms of

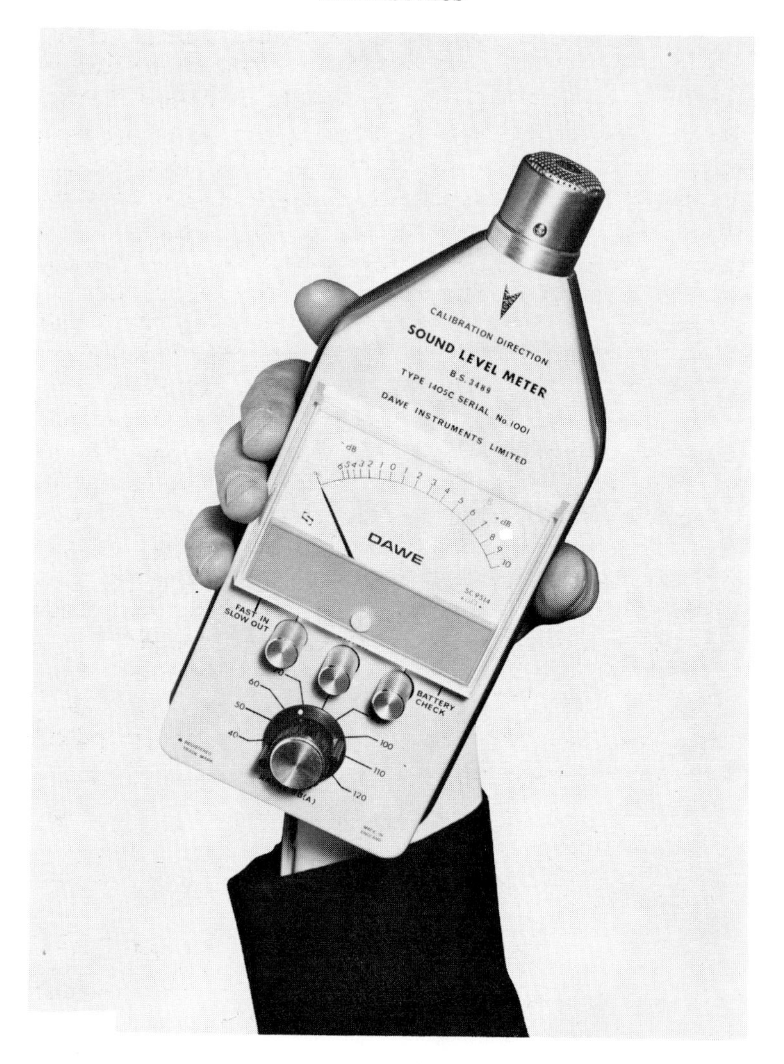

Fig. 3.3. Simple sound level meter. (Photograph by courtesy of Dawe Instruments Ltd.)

decibels which rely on a relative measurement of two intensities. In that case the intensity would be expressed in terms of a rate of flow of energy per unit area of cross section of the wave front. Indeed in scientific or technological work the intensity of ultrasound is generally measured in intensity units.

We must distinguish between loudness and intensity. The 'loudness' of a sound is a *subjective* measure that corresponds rather closely with the intensity, which is a physical quantity. Although loudness and intensity are closely related they are nevertheless distinct from one another. Loudness will depend on the frequency and also on the 'average person's response' to the physical intensity of the sound as well as on the intensity itself. We shall not devote any further attention to attempts to define a quantitative scale of loudness.

A sound-level meter, for measuring the intensity of an audible sound, consists of a microphone and an associated 'box' of electronics to convert the output from the microphone to give a reading of the intensity of the sound in decibels. Such instruments are generally portable (see fig. 3.3) and may easily be used 'on location' to measure 'noise pollution' due to road-traffic, railway trains, or aeroplanes and to measure noise levels in workshops and factories, etc. The threshold of 'hearing' for such an instrument will by no means necessarily coincide with the threshold of hearing for humans. A sound-level meter of this type could not be used to determine accurate values of attenuation coefficients. We shall see in Chapter 4 that considerable modifications in basic principles are needed to develop systems for detecting ultrasound, for measuring ultrasonic intensities, or for determining ultrasonic attenuation coefficients.

3.2. *Absorption in gases, liquids and solids*

The three important mechanisms that contribute to the absorption of sound or ultrasound in a gas are

(a) viscosity,
(b) thermal conduction, and
(c) thermal relaxation.

The attenuation coefficient α for a gas may be written as

$$\alpha = \alpha_{\text{vis}} + \alpha_{\text{th}} + \alpha_{\text{relax}} \tag{3.18}$$

where α_{vis}, α_{th}, and α_{relax} are the contributions arising from these mechanisms, respectively.

The first two of these absorption mechanisms can be described satisfactorily within the context of classical physics and the appropriate theory was developed in the nineteenth century. Sometimes one writes

$$\alpha_{\text{vis}} + \alpha_{\text{th}} = \alpha_{\text{class}}. \tag{3.19}$$

The energy in a wave of sound or ultrasound in a gas is the energy of the sinusoidal motions of the molecules in the path of the wave. In (a) and (b) the absorption involves converting these regular molecular motions into random (thermal) motions of the molecules and thereby removing energy from the original wave. Thermal conduction is involved because the compressions and rarefactions in the gas, in the path of the wave, are adiabatic (see also Section 2.1) rather than isothermal. Consequently temperature gradients will be set up between neighbouring regions of compression and rarefaction. Heat will flow between these regions and, because of the thermal resistance of the gas, some of the heat will be dissipated. This heat must be extracted from some source and that source is the energy of the ultrasonic beam; consequently the amplitude of the ultrasound will decrease as it travels through the gas.

To try to understand the third mechanism, thermal relaxation, we consider the kinetic energy of the molecules in a gas. In a monatomic gas all the thermal energy of the molecules will be in the form of translational kinetic energy of the molecules. For a polyatomic gas the thermal kinetic energy will be partly in the form of translational energy of the molecules, but it will also be partly in the form of internal energy of the molecules. This internal energy may be either kinetic energy of rotation of the molecule or it may be kinetic energy of vibration of the atoms in the molecule (see fig. 3.4). If energy is converted from the energy of a wave of sound or ultrasound, into random thermal energy of a polyatomic molecule, some of it will become random translational energy and some of it will become internal energy. It is the conversion into internal energy which is described as 'thermal relaxation'. We may expect α_{class} to account completely for the observed attenuation in monatomic gases, because these gases do not possess internal degrees of freedom (rotation and vibration) and so the thermal relaxation mechanism is not relevant. But for a gas with polyatomic molecules the thermal relaxation mechanism will be expected to be important.

Classical arguments were used in the middle of the nineteenth century to derive expressions for α_{vis} and α_{th}, namely

$$\alpha_{vis} = \frac{8\pi^3}{3}\left(\frac{\eta}{\rho c^3}\right)\nu^2 \tag{3.20}$$

and

$$\alpha_{th} = \left(\frac{2\pi^2 K}{\rho c^3 c_v}\right)\frac{(\gamma-1)}{\gamma}\nu^2 \tag{3.21}$$

where ν is the frequency and the remaining symbols have the following meanings:

ρ: density
η: coefficient of viscosity
c: speed of sound
K: coefficient of thermal conductivity
c_v: specific heat capacity at constant volume
γ: (c_p/c_v) (c_p = specific heat capacity at constant pressure).

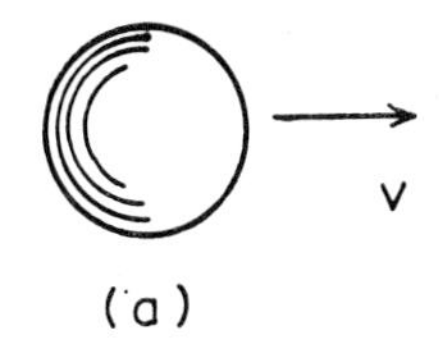

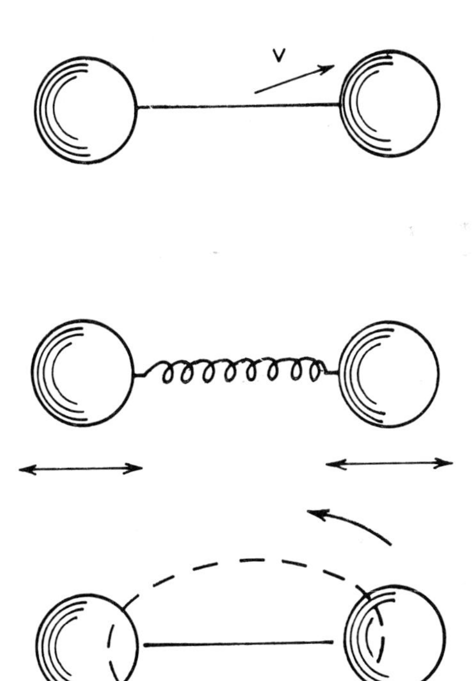

Fig. 3.4. Degrees of freedom. (*a*) Free atom: translation only. (*b*) Diatomic molecule: translation, vibration and rotation.

There is an upper limit to the frequency for which these formulae are valid although, in practice, they will be valid for all except the very highest ultrasonic frequencies. It is also possible to obtain expressions for α_{relax}, but the details are quite complicated; we simply note that, as with other mechanisms, the attenuation coefficient is proportional to ν^2 (again except when ν is extremely large).

Table 3.3. Calculated attention coefficients α_{class}.

	α_{class}/ν^2 10^{-11} Np m^{-1} s^2
Air	1.4
Carbon dioxide	1.3
Helium	0.52
Hydrogen	0.16
Neon	1.9
Nitrogen	2.4

(After J. Blitz, *Fundamentals of Ultrasonics* (Butterworths, London, 1967).)

Some values of α_{class} ($= \alpha_{vis} + \alpha_{th}$) calculated using eqns. (3.20) and (3.21) are given in Table 3.3. The measured values of the attenuation coefficient in air is also given in Table 3.1.

Ultrasound which is propagating through a liquid may be reduced in intensity by the three mechanisms that we noted for gases and also by a fourth mechanism, namely

(d) Structural relaxation.

For viscosity and thermal conduction the classical theory, which has already been mentioned for gases and which led to the results that we quoted in eqns. (3.20) and (3.21) for α_{vis} and α_{th}, can be applied to liquids as well as to gases. By substituting into eqns. (3.20) and (3.21) one finds that typical values of α_{vis} are of the order of $10^{-17}\nu^2$ Np cm^{-1}, and of α_{th} are of the order of $10^{-20}\nu^2$ Np cm^{-1}. Note that these are very much smaller than in a gas. Also the ratio α_{th}/α_{vis} is so small for a liquid that the thermal conduction mechanism can generally be ignored. However, for a liquid metal the value of K may be large enough to make α_{th} important. The measured attenuation in a liquid agrees well with eqn. (3.20) for a monatomic liquid, such as mercury or liquid argon. The measured attenuation also agrees with eqn. (3.20) for diatomic liquids, such as hydrogen, nitrogen, and oxygen, for which the temperatures are too low to excite any of the rotations or internal vibrations of the molecules, that is when the temperature is too low for thermal relaxation to contribute significantly to the attenuation.

In polyatomic liquids, as in polyatomic gases, the measured value of the attenuation is greater than that predicted by eqns. (3.20) and (3.21). For many liquids this discrepancy can be accounted for by thermal relaxation. This may involve the conversion of ultrasonic energy into energy of rotation or internal vibration of the molecules, as we have already noted

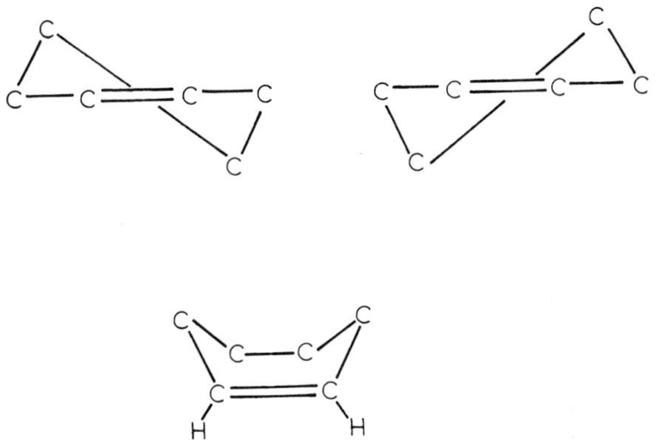

(a)

(b)

Fig. 3.5. (*a*) Half-chair forms and (*b*) half-boat form of cyclohexene; in each form four C atoms (two on either side of the double bond) are coplanar and most of the H atoms are omitted from the diagrams.

occurs for gases. However, it may also involve relaxation between two different isomeric forms of a molecule. For example, cyclohexene, C_6H_{10}, is a ring compound with one double bond; four of the six carbon atoms (and two of the hydrogen atoms) are coplanar. A cyclohexene molecule may exist in two isomeric forms, a 'half-chair' form and a 'half-boat' form.* There are actually two equivalent half-chair forms (see fig. 3.5). The half-chair forms are more stable than the half-boat form by about 11 kJ per mole. Thermal relaxation between isomeric forms, as a mechanism for ultrasonic absorption in such a liquid, involves individual molecules undergoing transitions from the form with lower energy to that with higher energy, at the expense of the energy of the ultrasonic beam. This mechanism involves the rearrangement of the structures of the individual molecules. There is also another possibility, called 'structural relaxation', which can be observed in associated liquids, that is in liquids in which the molecules are held together in groups by the intermolecular forces. In this case the ultrasonic beam loses energy to cause additional rearrangements of the groups of associated molecules, beyond the rearrangements that ordinarily occur as a result of thermal fluctuations.

Since thc attenuation of ultrasound in liquids is very much smaller than in gases, at least at rather low ultrasonic frequencies, ultrasound can be used under water in detection and ranging ('sonar') devices over very large distances indeed (see Section 5.1.).

*This nomenclature is derived from that used for cyclohexane which occurs in a 'chair' form and a 'boat' form (and also in a flexible form).

We have just seen that for a monatomic gas or liquid it is possible to achieve a considerable degree of quantitative agreement between the theory and experiment of ultrasonic attenuation, α_{class}, by using classical arguments which were, essentially, based on the continuum model. It was only for polyatomic gases and liquids that it became important to adopt a molecular viewpoint in order to obtain even a qualitative understanding of the absorption mechanisms. But for solids it is convenient to adopt a molecular viewpoint at the very start. Possible attenuation mechanisms, for a single-crystal specimen, include the following:

(*a*) *The thermoelastic effect.* This is similar to the attenuation due to thermal conduction which we have described for gases and liquids. But the formula quoted for α_{th} does not apply to solids; it is necessary to use more complicated formulae.

(*b*) *Scattering by the thermal vibrations of the medium.* We have already seen in Section 2.4 that it is useful to regard the thermal vibrations of a solid as being equivalent to a set of sound waves or ultrasonic waves of many different frequencies propagating randomly in all directions in the material. In this scattering mechanism there is an interaction between a beam of ultrasound and these random waves of sound or ultrasound leading to a scattering of some of the energy in the beam. We can adopt the quantum-mechanical or 'quasi-particle' viewpoint and regard an ultrasonic wave as being quantized and consisting of a flow of particle-like *phonons*. In this scattering mechanism we then have some of the phonons which were travelling in a special direction in the ultrasonic beam, being scattered by interacting with some of the random phonons in the thermal vibrations. This would then, alternatively, be described as phonon-phonon scattering.

The attenuation due to scattering by the thermal vibrations of the medium is strongly temperature-dependent (see fig. 3.6). This is important in isolating this contribution, experimentally, from other contributions to the attenuation.

(*c*) *Dislocation damping.* Suppose that an external stress is applied to a crystal that contains dislocations. In addition to the elastic strain which would develop in the crystal in the absence of the dislocations, there will be an additional strain caused by displacements of the dislocations. Thus, as an ultrasonic wave travels through a crystal that contains dislocations there will be changes in the dislocation strain as the ultrasound unpins and moves the dislocation network. The attenuation due to this mechanism passes through a maximum at frequencies in the upper kHz to lower

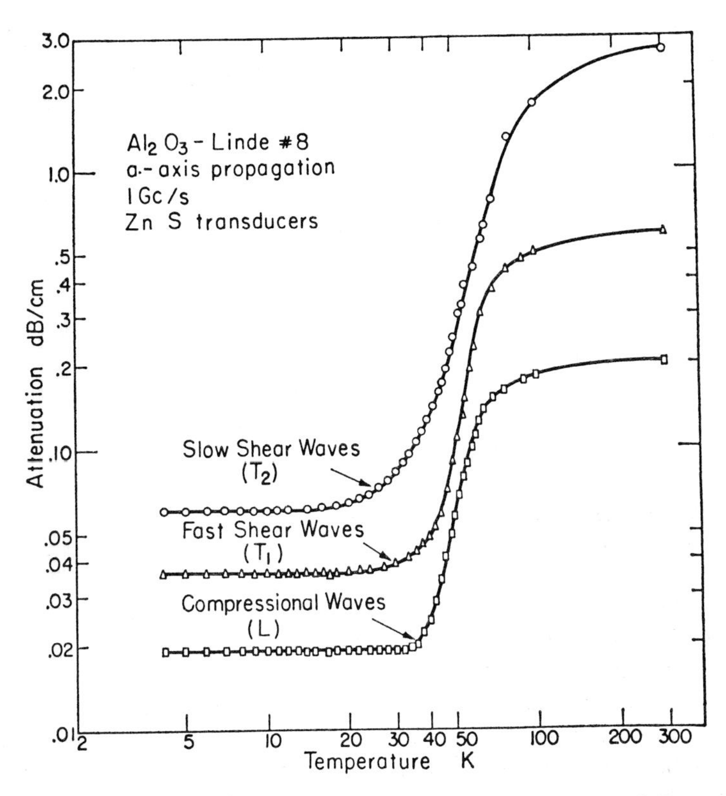

Fig. 3.6. Temperature-dependent ultrasonic attenuation in Al_2O_3 attributed to phonon–phonon interactions (From J. De Klerk, 1965, *Phys. Rev.* **139**, A1635.)

MHz range; it facilitates the estimation of the dislocation density and loop length. Damage due to irradiation and to 'working' can be investigated at lower defect concentrations than are possible using optical absorption techniques.

(*d*) *Magnetoelastic interactions.* In a magnetically-ordered material an ultrasonic beam may be attenuated by interaction with domain walls. Within a single domain it may also be attenuated by interaction with the thermal fluctuations of the orientations of the magnetic moments. In the same way that the thermal vibrations of a solid can usefully be regarded as being equivalent to a set of waves, so also the thermal vibrations of the magnetic moments in a magnetically-ordered material can be regarded as being equivalent to a set of waves, called *spin waves*, propagating randomly in all directions in the material. Thus the interaction between an ultrasonic beam and the thermal fluctuations of the magnetic moments can be regarded as an interaction between ultrasonic waves and spin waves. In a similar way, ultrasound may also be attenuated by interaction with thermal fluctuations in the spontaneous electric polarization in a ferroelectric material.

(e) *Interaction with conduction electrons in a metal.* In a metal energy from an ultrasonic wave is transferred to the conduction electrons when they are accelerated by the electric fields induced by the wave's displacement of lattice ions (the electron-phonon interaction). The mean free paths of the electrons must be longer than the wavelength of the ultrasound for an appreciable absorption of energy to occur; consequently, this mechanism is only important at low temperatures. This contribution, which in the MHz range yields attenuation proportional to ν^2, can be isolated by its variation as the paths of the electrons are altered by an applied magnetic field or by its behaviour if the metal becomes superconducting (see also Section 9.3).

To all these mechanisms, which apply to pure single-crystal specimens, we may add for a polycrystalline specimen or a disordered alloy,

(f) *Scattering by inhomogeneities in the material.* This is the principal cause of loss of energy in the ultrasonic testing of materials. The inhomogeneities may consist of different grain orientations in a polycrystalline substance, since each individual grain is anisotropic for the propagation of ultrasound, or of inclusions of different materials, or of porosity as in sintered materials. This scattering is significant for inhomogeneities larger than about one-hundredth of the ultrasonic wavelength. The scattering increases as the third power of the grain size and the fourth power of the frequency as long as the wavelength remains greater than the grain size (Rayleigh scattering). When the ultrasonic wavelength becomes comparable with the grain separation, the scattering centres cease to act independently and the ultrasonic velocity is affected.

If one seeks to turn the above discussion of ultrasonic attenuation in solids into quantitative terms, one is faced with a problem that is much greater than for ultrasonic attenuation in gases and liquids. For a solid it is difficult, although not impossible, to calculate *ab initio* theoretical values for each of the various contributions to $\alpha_I(\nu)$ for a given material. It also has to be remembered, of course, that in measuring the attenuation of a beam of ultrasound in a material, one is measuring the total attenuation arising from all the different mechanisms involved. It may not be possible to devise an experiment that will isolate the attenuation due to just one of these mechanisms. Nevertheless, ultrasonic attenuation has proved very valuable in a number of different experimental techniques which are used in basic research in solid state physics; we shall mention some of these in Chapter 9.

Before concluding this section we ought to mention the idea of a *relaxation time*, usually denoted by τ. This is a general concept which is appropriate to any absorption mechanism and which can be applied either to classical or to quantum-mechanical descriptions of an absorption process. We shall use the quantum-mechanical approach. Suppose that an ultrasonic beam encounters some system, such as a molecule of a gas, which usually exists in a ground state, with energy E_0. Suppose also that if ultrasound is absorbed this system will be raised to its first excited state, with energy E_1. The excited system will then return to its ground state E_0, and will lose an amount of energy $(E_1 - E_0)$ in some way or another; the destination of this energy is not of great importance to us just now. This 'decay' from E_1 to E_0 is a random process and the relaxation time τ is a (quantum-mechanical) average of the time that the system can be expected to survive in its excited state E_1. Alternatively, the probability that the system will have returned from E_1 to E_0, at a time t after the initial absorption of energy, can be written as $[1 - \exp(-t/\tau)]$. One can consider separately each of the absorption mechanisms that may contribute to the absorption of an ultrasonic wave and can define an appropriate relaxation time for each of them. Thus, for a gas we would have τ_{vis}, τ_{th}, and τ_{relax} for the absorption due to viscosity, thermal conduction, and thermal relaxation, respectively. Although we shall not make any very great use of the idea of a relaxation time in this book, it is nevertheless an important concept that is widely used in more advanced discussions of absorption mechanisms.

3.3. *Sensing and control applications*

Suppose that a generator of ultrasound at A produces a beam of ultrasound which is then received by a detector at B (fig. 3.7(*a*)). If an object is placed in the way of the beam, the ultrasound will no longer reach the detector (fig. 3.7(*b*)), and the resulting response of the detector can then be used to activate some piece of electrical or mechanical machinery. The reader may be more familiar with the use of a beam of light and a photocell than with the ultrasonic counterpart. Either the ultrasonic method or the optical method may have some advantage over the other; which one is better for any given application depends on that application.

One obvious application would be to detect a person entering a room. This could be used, instead of an optical beam and a photocell, to control

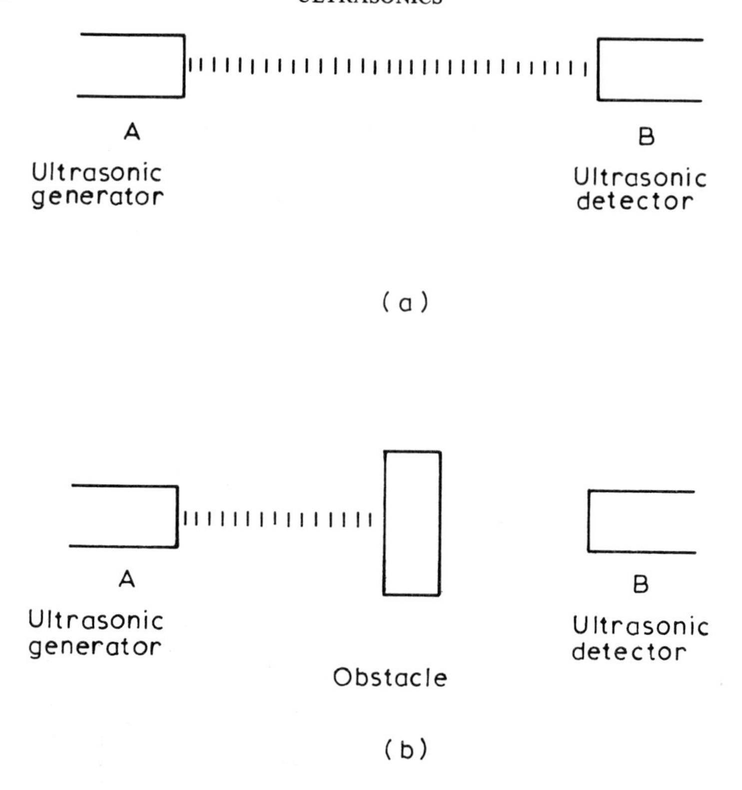

Fig. 3.7. The obstruction of an ultrasonic beam by an obstacle.

the opening and closing of a door, or to operate a counter to count the number of people entering a room. Ultrasonic beams and detectors can be used to operate burglar alarms, and to operate counters for the counting of artefacts on a factory production line.

An ultrasonic beam can also be used in more sophisticated measuring or control applications, such as positioning, size control, level control, and load-height control, as well as, of course, the automatic opening and closing of doors (fig. 3.8). Another interesting possible application is of a 'foolproof' ultrasonic seat-belt which was developed by the Ford Motor Company (fig. 3.9). In a car containing this device a sensor in the seat (C) is depressed by the weight of the driver who must buckle the belt properly across his body (B). Only then is an ultrasonic signal emitted from a small transmitter mounted on the belt (A) to a receiver on the pillar to complete the electronic circuit before the car's engine can be started.

The control applications mentioned so far are essentially 'on-off'

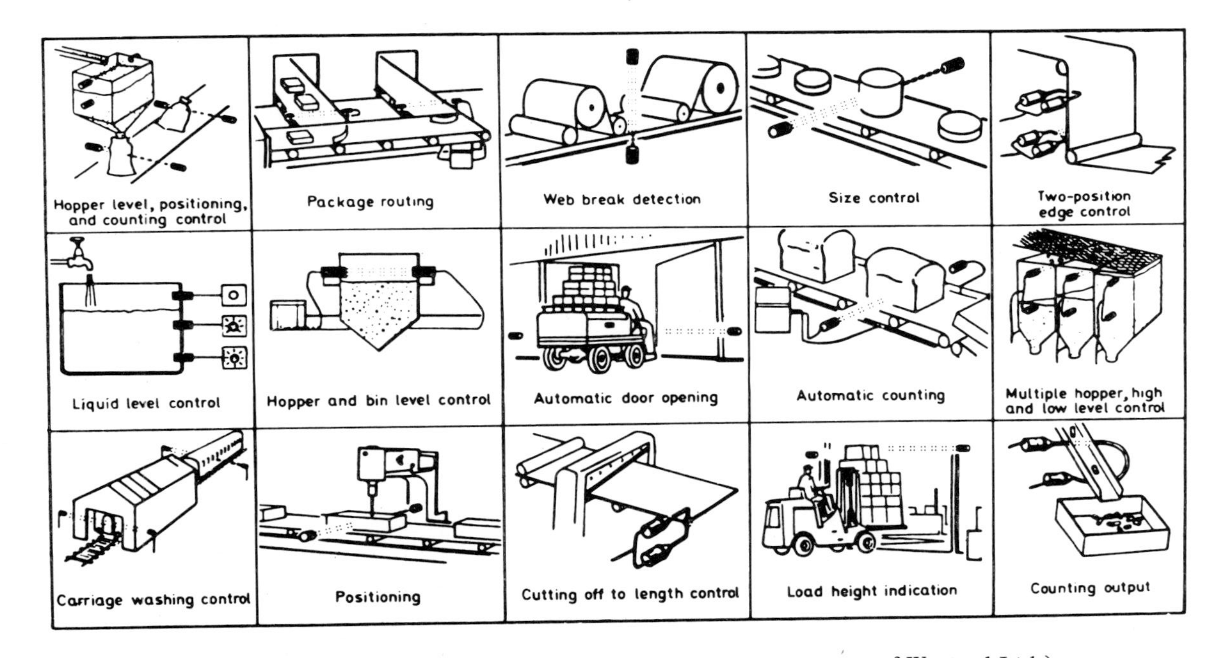

Fig. 3.8. Various ultrasonic sensing applications (Diagram by courtesy of Westool Ltd.)

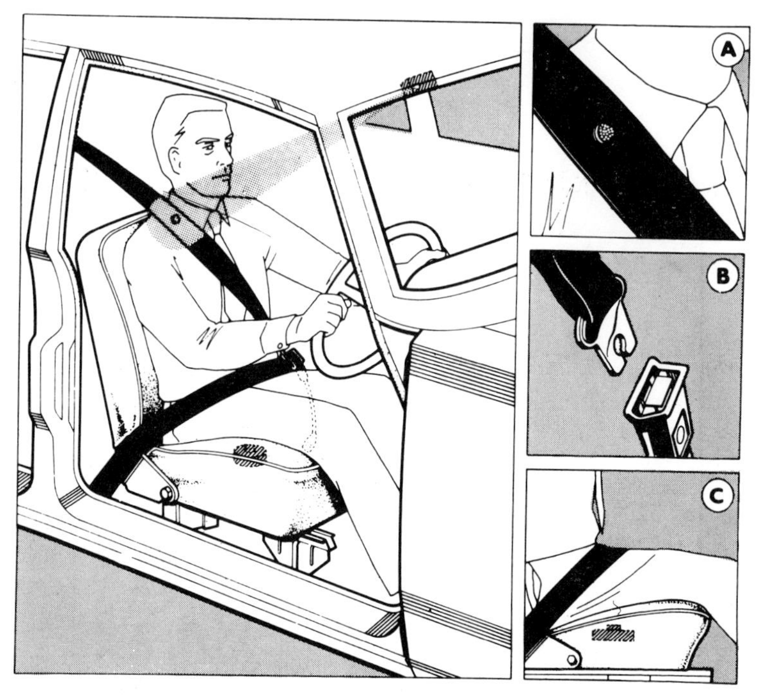

Fig. 3.9. 'Foolproof' ultrasonic seat belt. (Photograph by courtesy of Ford Motor Co. Ltd.)

switches, but more sophisticated systems can be constructed to monitor some parameter which is a continuous variable. Thus, for example, a system has been developed by Sliwiński and Walasiak for controlling the concentration of a transparent solution. This system is based on the fact that the speed, and hence the wavelength, for ultrasound of a given frequency will depend on the concentration of a solution; the wavelength is monitored by using the ultrasound to form a diffraction grating as in the method of Debye and Sears which will be described later (see Section 7.1).

4. The generation and detection of ultrasound

4.1. *Mechanical generators and detectors*

The early nineteenth century work on ultrasound used mechanical generators which were developments of whistles, sirens, or tuning forks. For example, in 1883 Galton described the use of a whistle made from a brass tube with an internal diameter of less than one-tenth of an inch and with a movable plug at one end to vary the effective length of the whistle. A development of Galton's whistles, due to Edelmann in 1900, is illustrated in fig. 4.1. Compressed air enters at A and leaves through an annular orifice at D to impinge on a circular knife edge at one end of the resonant cavity E. The air gap is adjusted at B, C, and the resonant length of the cavity (V) is altered by a plunger (S) attached to F, G.

It is possible to estimate the upper limit to the frequencies that can be obtained with ultrasonic whistles. If we regard the whistle as a resonant cavity of length l, which is open at one end and closed at the other, the wavelengths of the fundamental and of the first few harmonics are given by (see fig. 4.2)

$$l + \delta = \frac{\lambda}{4}, \frac{3\lambda}{4}, \frac{5\lambda}{4}, \ldots \text{etc.} \tag{4.1}$$

where δ is the end correction of the tube.

The corresponding frequencies are given by

$$\nu = \frac{c}{4(l+\delta)}, \frac{3c}{4(l+\delta)}, \frac{5c}{4(l+\delta)}, \ldots \text{etc.} \tag{4.2}$$

If the ultrasonic energy is mainly concentrated in the fundamental, we can use

$$\nu = \frac{c}{4(l+\delta)} \tag{4.3}$$

to calculate values of ν corresponding to various values of l. Equation (4.3) will cease to be valid if the length of the cavity becomes too small

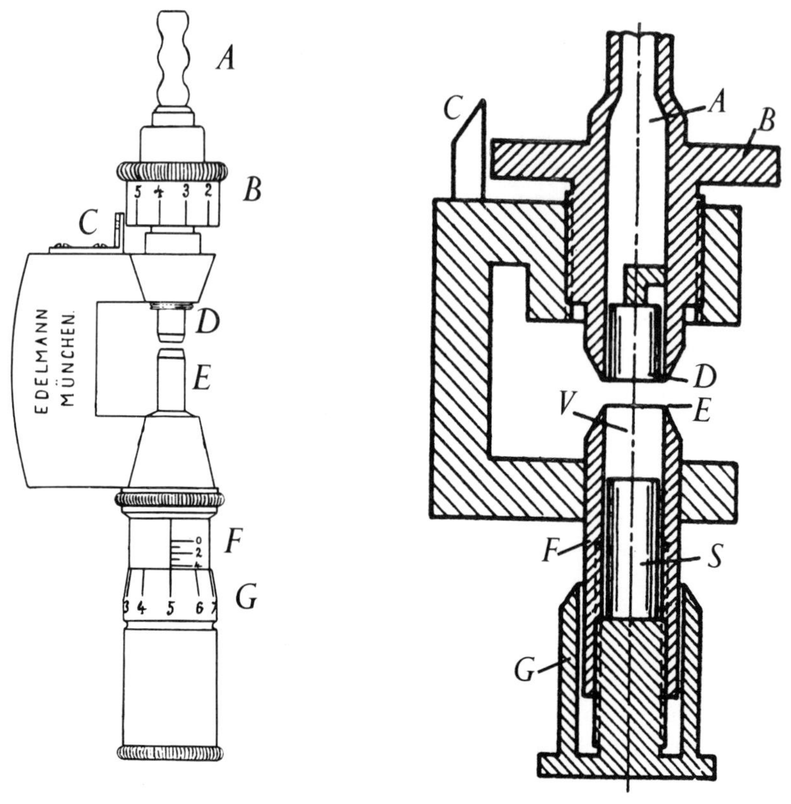

Fig. 4.1. (a) A development of Galton's whistle due to Edelmann in 1900, (b) section through whistle illustrated in (a).

compared with its diameter; as a rough and ready rule one can say that the length should be at least $1\frac{1}{2}$ times the diameter. To produce a high frequency we need a small value of l and this in turn requires the diameter of the tube to be even smaller. If we take the smallest reasonable value of $(l + \delta)$ to be of the order of 2 or 3 mm and use the value of $c = 330\,\text{m}\,\text{s}^{-1}$ for air, we find the upper limit for ν, using eqn. (4.3), to be about 30–40 kHz.

Sirens may also be used as generators of sound or ultrasound (fig. 4.3). The jet of air is interrupted every time the blank part of the disc between adjacent holes passes in front of the nozzle, so that the frequency generated is given by

$$\nu = Nn \tag{4.4}$$

where N is the number of holes in the disc and n is the number of rotations of the disc per second. Instead of a single nozzle one may use a second

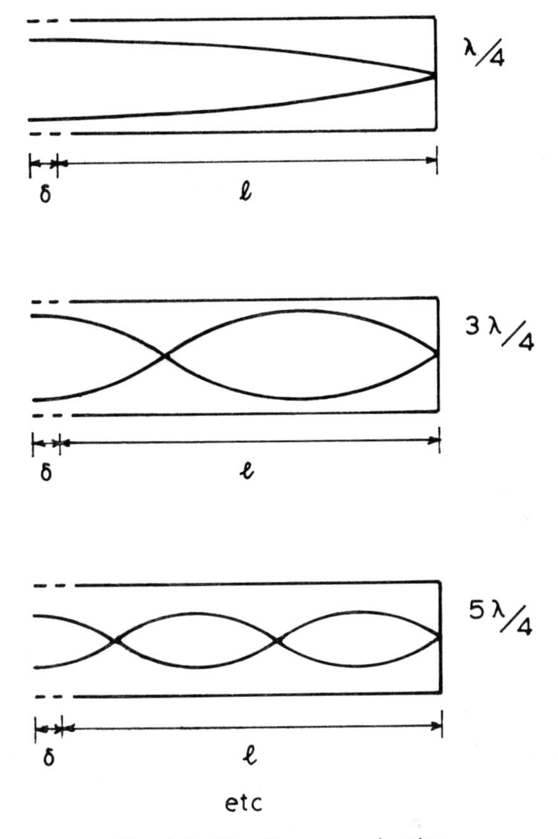

Fig. 4.2. Standing waves in pipe.

stationary disc (a stator) with the same configuration of holes as in the rotating disc. A siren developed for generating high energy ultrasound is illustrated in Fig. 4.4. The stator and the rotor each have 100 holes and the rotor can be rotated at up to 340 revolutions per second, giving an upper limit of 34 kHz for the frequency.

Although mechanical ultrasonic generators are not widely used nowadays, there are certain applications for which they are particularly suitable (see, for example, Section 8.3.).

Mechanical ultrasonic generators are limited to rather low frequencies, at least by modern ultrasonic standards. Nevertheless it was possible, even in the early days, to calibrate them reasonably accurately with regard to frequency. The situation was much less satisfactory with regard to the early detectors. One can use the ear as a 'go/no-go' detector provided that the frequency of the sound is below the limit of hearing for

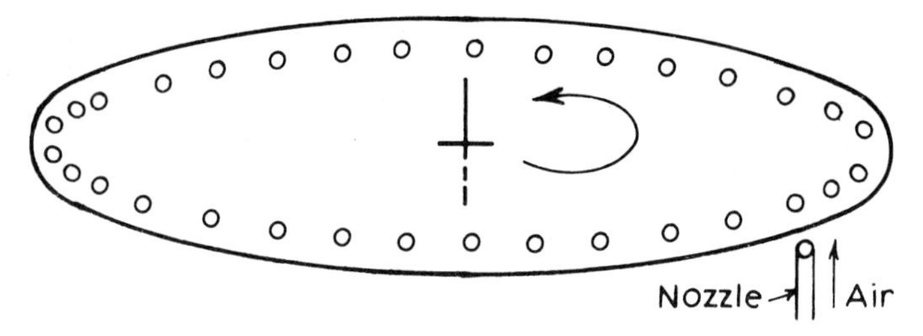

Fig. 4.3. The principle of the siren.

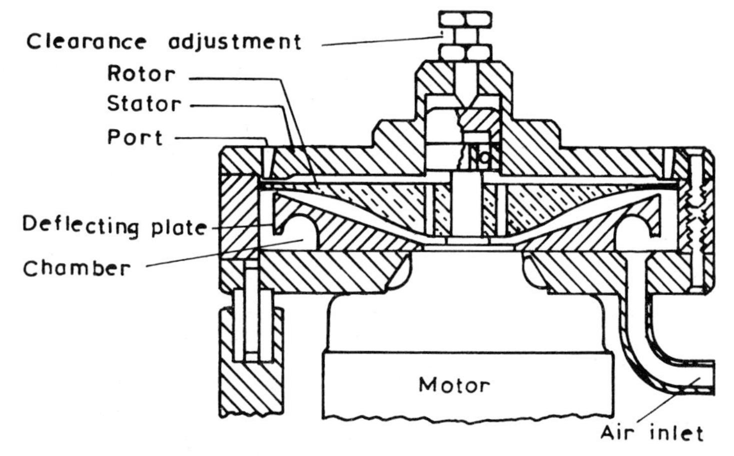

Fig. 4.4. Ultrasonic siren (From C.H. Allen and I. Rudnick, *J. Acoust. Soc. Amer.*, 1947, **19**, 857.)

that ear, but this is not very satisfactory. We quote from Galton's work again,* this time on animals.

> I have tried experiments with all kinds of animals on their powers of hearing shrill notes. I have gone through the whole of the Zoological Gardens, using an apparatus arranged for the purpose. It consists of one of my little whistles at the end of a walking-stick—that is, in reality, a long tube; it has a bit of india-rubber pipe under the handle, a sudden squeeze upon which forces a little air into the whistle and causes it to sound. I hold it as near as is safe to the ears of the animals, and when they are quite accustomed to its presence and heedless of it, I make it sound; then if they prick their ears it shows that they hear the whistle; if they do not, it is probably inaudible to them. Still, it is very possible that in some cases they hear but do not heed the sound. Of all creatures, I have found none superior to cats in the power of hearing shrill sounds; it is perfectly remarkable what a faculty

*F. Galton, *Inquiries into Human Faculty and Development* (MacMillan, 1883).

they have in this way. Cats, of course, have to deal in the dark with mice, and to find them out by their squealing. Many people cannot hear the shrill squeal of a mouse. Some time ago, singing mice were exhibited in London, and of the people who went to hear them, some could hear nothing, whilst others could hear a little, and others again could hear much. Cats are differentiated by natural selection until they have a power of hearing all the high notes made by mice and other little creatures that they have to catch.

For a long time all that science had to offer for the detection of ultrasound, or to replace the ear as a detector of sound at audible frequencies, was the 'sensitive flame'. For a description of the basic phenomena of sensitive flames we quote from Lord Rayleigh.*

The earliest observation upon this subject was that of Professor Leconte, who noticed the jumping of the flame from an ordinary fishtail burner in response to certain notes of a violoncello. The sensitive condition demanded that in the absence of sound the flame should be on the point of flaring. When the pressure of gas was reduced, the sensitiveness was lost.

An independent observation of the same nature drew the attention of Professor Barrett to sensitive flames; and he investigated the kind of burner best suited to work with the ordinary pressure of the gas mains. 'It is formed of glass tubing about $\frac{3}{8}$ of an inch (1 cm) in diameter, contracted to an orifice $\frac{1}{16}$ of an inch (0·16 cm) in diameter. It is very essential that this orifice should be slightly V-shaped . . . '

But the most striking by far is the high-pressure flame employed by Tyndall. The gas is supplied from a special holder under a pressure of say 25 cm of water to a pinhole steatite burner, and the flame rises to a height of about 40 cm. Under the influence of a sound of suitable (very high) pitch the flame roars, and drops down to perhaps half its original height. Tyndall showed that the seat of sensitiveness is at the root of the flame. Sound coming along a tube is ineffective when presented to the flame a little higher up, and also when caused to impinge upon the burner below the place of issue.

It is to Tyndall that we owe also the demonstration that it is not to the flame as such that these extraordinary effects are to be ascribed. Phenomena substantially the same are obtained when a jet of unignited gas, or carbonic acid,† hydrogen, or even air itself, issues from an orifice under proper pressure. They may be rendered visible in two ways. By association with smoke the whole course of the jet may be made apparent; and it is found that suitable smoke jets can surpass even flames in delicacy. 'The notes here effective are of much lower pitch than those which are most efficient in the case of flames'. Another way of making the sensitiveness of an air-jet visible to the eye is to cause it to impinge upon a flame, such as a candle flame, which plays merely the part of an indicator.

*J.W. Strutt, Baron Rayleigh, *The Theory of Sound*, 2nd ed. (Macmillan, 1894, reprinted by Dover Books, 1945).

†CO_2 gas.

The theory of the behaviour of a sensitive flame in the presence of ultrasound is far from trivial. In view of the fact that the use of such flames is largely obsolete anyway now, there is little point in attempting to pursue that theory here. While such flames act as detectors of ultrasound they can, at best, give only qualitative information about either the frequency or the intensity of the ultrasound. A number of mechanical detectors of ultrasound have been invented over the years; while some of them may be used at the present time for some specialized purposes, their use is not generally widespread.

4.2. *Piezoelectric transducers*

We have seen that it is impossible to devise mechanical oscillators to operate at extremely high frequencies. Therefore, for very high frequency ultrasonic work it is necessary to find an alternative means of generating ultrasound. This is now most commonly done by generating electrical oscillations of the required frequency and then converting them into mechanical oscillations. In this way it is possible to define the frequency quite accurately. Moreover, the use of electrical oscillations, which is necessary for the generation of very high frequency ultrasound, is often also convenient in practice for the generation of ultrasound at lower frequencies too. Likewise we have noted that it is difficult to detect ultrasound, let alone measure ultrasonic intensities, if one attempts to do this mechanically; consequently, nowadays one converts the mechanical oscillations back into electrical oscillations. When interconverting electrical and ultrasonic oscillations the active components are referred to as *transducers*. By far the most common way of performing the electromechanical conversions in the generation and detection of ultrasound is to make use of the property of piezoelectricity.

Suppose that a slice of an electrically insulating crystal is compressed by applying forces to the surfaces of the slice. The crystal will be slightly deformed by this and, if certain conditions are satisfied by the symmetry of the internal structure of the crystal (see Further Reading for references), electric charges may appear on the slice (fig. 4.5), positive charges on one surface and negative charges on the opposite one. An electrostatic potential difference is thereby produced between the surfaces and an electrostatic field is created within the crystal. If the external forces are removed the charges disappear. A crystal which exhibits this phenomenon is said to be *piezoelectric*.

In the converse of the piezoelectric effect we suppose that a piezoelectric crystal is placed in an electric field and the crystal will then develop some distortion; if the electric field is removed the distortion will disappear. A convenient way to produce the electric field is to coat the

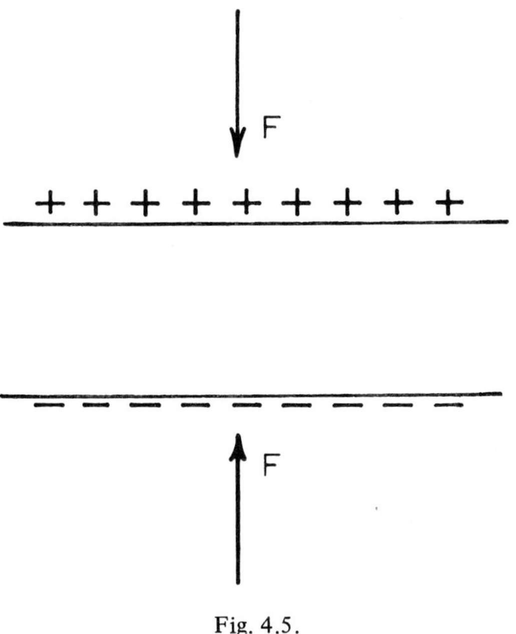

Fig. 4.5.

surfaces of the plate with a conducting film and to apply a voltage between these metal coatings.

The existence of the piezoelectric property depends on the internal arrangement of the atoms within the crystal. There are many crystals which, when they are subjected to a stress (fig. 4.5), undergo distortions in which the positive and negative charges contrive necessarily, because of the symmetry of the crystal, to arrange themselves in such a manner that no dipole moment arises in the specimen. The study of crystal symmetries, with a view to identifying those crystals which are 'allowed' to exhibit piezoelectricity and those which are 'forbidden' to exhibit piezoelectricity, is a specialized topic that is outside the scope of this book.

The piezoelectric behaviour of a given specimen depends both on internal symmetry of the crystal and on the orientation of the chosen crystal slice relative to the crystallographic axes. We shall describe what happens in the case of quartz, which occurs naturally in large single crystals. The piezoelectric properties of materials with other symmetries will differ from those of quartz.

The principal axis of symmetry of a crystal of quartz (fig. 4.6) is a three-fold axis of symmetry, as z in fig. 4.6. If the crystal is rotated clockwise or anticlockwise through $2\pi/3$ about the z axis its new appear-

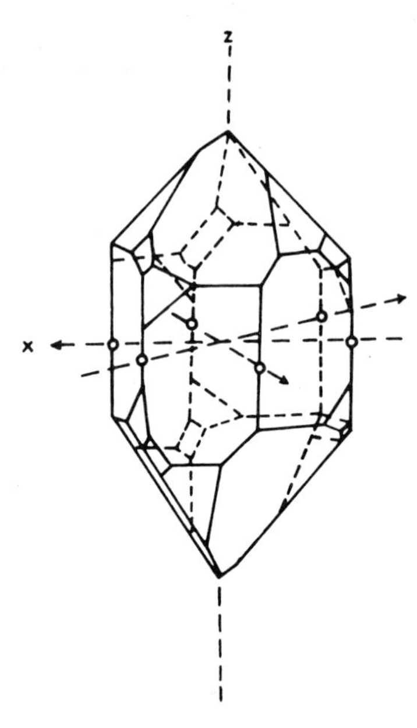

Fig. 4.6. Sketch of possible quartz crystal.

ance will be indistinguishable from that before rotation was performed. In a similar manner the axis x is a two-fold axis of symmetry. The y axis is at right angles to both x and z. Suppose now that we cut a rectangular block of quartz from such a crystal so that the edges of the block are parallel to the x, y, and z axes (see fig. 4.7). We shall use the labels '1', '2' and '3' for the pairs of faces normal to the x, y, and z axes, that is

pair of faces	label
OPQR, STUV	1
ORVS, PQUT	2
OPTS, RQUV	3.

In manufacturing transducers one uses thin slices rather than rectangular blocks of quartz. A slice which is cut so that its large surfaces are normal to an x axis are called 'x-cut' slices and slices with their large surfaces normal to a y axis are called 'y-cut' slices.

Consider the pairs of faces '1' which are normal to the x axis. One can apply stress either perpendicular or parallel to these faces; in the former case (T_{11} in fig. 4.7) we have a tensile or compressional stress and in the latter case (T_{21} and T_{31} in fig. 4.7) we have shear stresses. The tensile or compressional stress T_{11}, i.e. on faces '1', causes charges to appear on the faces labelled '1' but on no other faces. A tensile stress

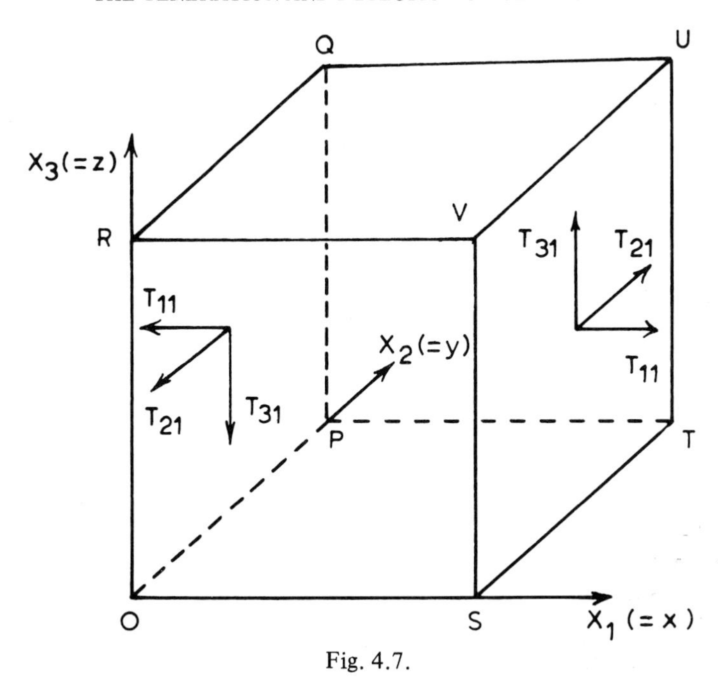

Fig. 4.7.

applied parallel to y, i.e. on faces '2' also causes charges to appear on the faces labelled '1' (not '2') and on no other faces. A tensile stress applied parallel to z, i.e. on the faces '3' produces no charges at all on any of the faces. This means that a plate of quartz which is cut with its faces perpendicular to the [0001] axis (the z axis) will not develop any charges on its faces if a compression or tension is applied normal to these faces. Thus, whereas there are many crystals which cannot exhibit piezoelectricity at all, even for a material which is in principle allowed to be piezoelectric, it may still be possible to cut a sample in such a way that the piezoelectric behaviour is suppressed. The piezoelectric properties of the block of quartz sketched in fig. 4.7 for tensile or compressional stresses can be summarized as follows:

Direction of tensile force	*Charges appear on face*
along x axis	'1'
along y axis	'1'
along z axis	no charges at all.

There are altogether six possible shear stresses (in each of two possible directions acting on each of three possible pairs of faces). Of these six shear stresses, there are two which produce charges on the faces '1' and on no other faces while the remaining four all produce charges on the faces '2' and on no other faces. This exhausts all the possible kinds of stress that can be applied to this rectangular block. There is no stress at all which can produce any charges on the pair of faces labelled '3', that is on the pair of 'horizontal' faces which are normal to the three-fold axis of the quartz. This description so far has been in terms of various applied stresses causing consequential charge distributions. It can now be inverted directly to find the stresses which will appear when a voltage is applied to plates fixed to any given pair of opposite faces of the rectangular block. Different and more complicated, piezoelectric behaviour would occur for a quartz block cut in a different orientation with respect to the crystallographic axes, or for a piezoelectric crystal with point-group symmetry different from that of quartz.

What we have said so far in this section might be taken to imply that the piezoelectric effect only occurs in slices or blocks of well-formed single-crystalline material. This is not the case and it is also possible to observe piezoelectricity in certain polycrystalline materials or amorphous materials.

The importance of the piezoelectric effect and of the inverse effect in connection with ultrasound is that they provide a means of converting electrical oscillations into mechanical oscillations and *vice versa*, using piezoelectric transducers. Suppose that electrodes are fixed to the surfaces of an 'x-cut' rectangular slice of quartz. One electrode may be cemented or vacuum-deposited onto the back face of the transducer; for the other electrode, earthing the test piece avoids the problem of having wires at the transmitting (or receiving) face of the transducer. If an oscillating voltage of frequency ν is applied to these electrodes, mechanical oscillations of rather small amplitude will occur. These oscillations will have components in both the x and y directions, but not in the z direction.

Historically it was a long time before people fully appreciated that by far the most efficient electromechanical conversion will be achieved if the frequency of the electrical oscillations is equal to one of the natural mechanical resonance frequencies of the specimen (fig. 4.8). By suitable choice of the operating frequency and of the orientation of the slice, relative to the crystallographic axes of the quartz, it is possible to generate (almost pure) longitudinal ultrasonic waves or (almost pure) transverse waves with a given polarization.

It is perhaps important to mention that in using a piezoelectric transducer, or any other transducer, as an ultrasonic generator one does not usually take a separate variable-frequency oscillator (or signal generator) and apply its output to the transducer. The transducer is actually part of the oscillator circuit itself, and it is the chosen resonance frequency of the crystal which stabilizes the frequency of the electrical oscillations;

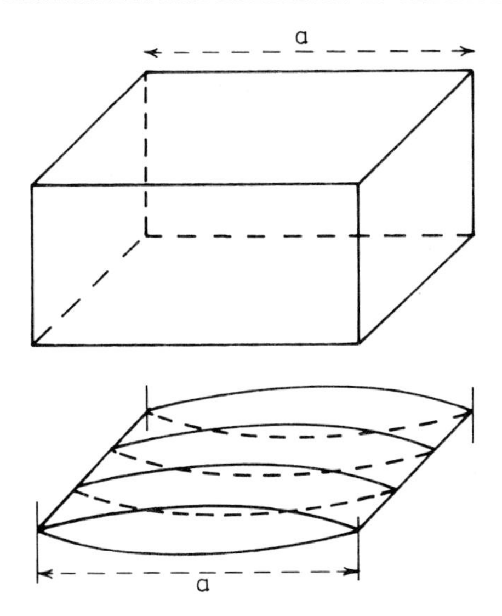

Fig. 4.8. Resonance condition for fundamental resonance for one linear dimension of a rectangular block clamped at its ends.

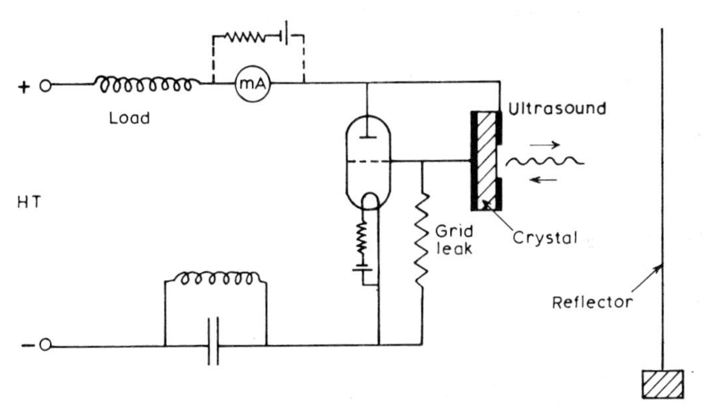

Fig. 4.9. One of the circuits used by Pierce in 1925 for crystal oscillator as a source of ultrasound.

such a circuit used by Pierce in 1925 is illustrated in fig. 4.9. When used as a receiver for ultrasound, the voltage developed between a pair of electrodes on opposite surfaces of a piezoelectric transducer is amplified and processed by a suitable electronic system to record or display the signal in a convenient manner.

For constant-frequency work the operating frequency would be chosen to be one of the natural resonance frequencies of the crystal. Below about

Table 4.1. Some examples of quartz transducer dimensions and resonance frequencies.

Frequency MHz	Quartz transducer thickness mm		Corresponding wavelengths mm			
			Steel		Water	Air
	x-cut	y-cut	bulk waves	shear waves		
1	2.88	1.92	6	3	1.5	0.3
10	0.288	0.192	0.6	0.3	0.15	0.03
	Corresponding velocities m s^{-1}					
	5760	3750	5900	3200	1500	331

10 MHz one would usually excite the fundamental frequency of a quartz plate and would choose its dimensions accordingly (see Table 4.1). For very low frequencies one can use either a damped quartz plate below its natural resonance frequency or a magnetostrictive transducer (see Section 4.3). If it is desired to sweep the frequency of the ultrasound, for example to avoid setting up stationary waves in the sample being studied, the transducer plate must be heavily damped to broaden the resonance; also if short pulses are to be transmitted a bandwidth of the order of 1/(pulse duration) is required. Suitable damping can often be achieved with a vulcanized rubber backing cemented to the transducer.

For fixed frequencies above about 10 MHz one would use higher harmonics so as to avoid having very thin quartz plates. Of the higher harmonics, only the *odd* ones can be generated. If we regard the specimen of piezoelectric material as the dielectric in a parallel-plate capacitor the even harmonics correspond to induced charges of the same sign on opposite surfaces of the specimen (fig. 4.10), so they will not be excited by a potential difference applied across it. There is a limit to the frequency of ultrasound that can be generated using higher harmonics. Extremely high frequencies in the higher microwave ultrasonic range, require other special kinds of transducer excited using microwave techniques which differ very much from those of conventional ultrasonics. These very high frequencies are not used in most of the everyday applications of ultrasound and are mainly of specialized interest (see Chapter 9).

There are many other piezoelectric materials besides quartz. Two which are commonly used in transducers are barium titanate, $BaTiO_3$, and lead niobate, $PbNb_2O_6$. These two materials are ferroelectric as well as being

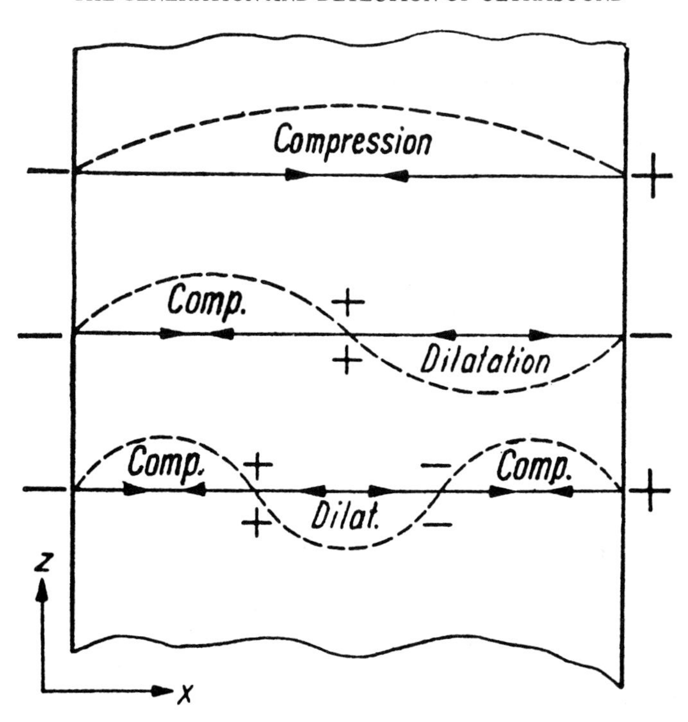

Fig. 4.10. Pressure and charge distribution in a quartz plate vibrating in (*a*) the fundamental, (*b*) second harmonic and (*c*) third harmonic (From L. Bergmann, *Ultrasonics and Their Scientific and Technical Applications* Wiley, New York, 1943.)

piezoelectric. That is, they are spontaneously polarized and a mechanical deformation causes a change in this polarization. Barium titanate and lead niobate cannot be obtained as large single crystals and so transducer material is prepared as a ceramic, by grinding the available crystals with binders and sintering, under pressure above 1000°C to form a ceramic. The crystallites are then aligned by cooling the material from above the ferroelectric transition temperature in an applied electric field; the reason for choosing materials which are not only piezoelectric but also ferroelectric is to make it possible to use this method of alignment.

4.3. *Other transducers*

There are also many other possible types of transducer which may have advantages in particular applications; these include magnetostrictive, electromagnetic, and electrostatic transducers.

Magnetostriction occurs for ferromagnetic and ferrimagnetic materials;

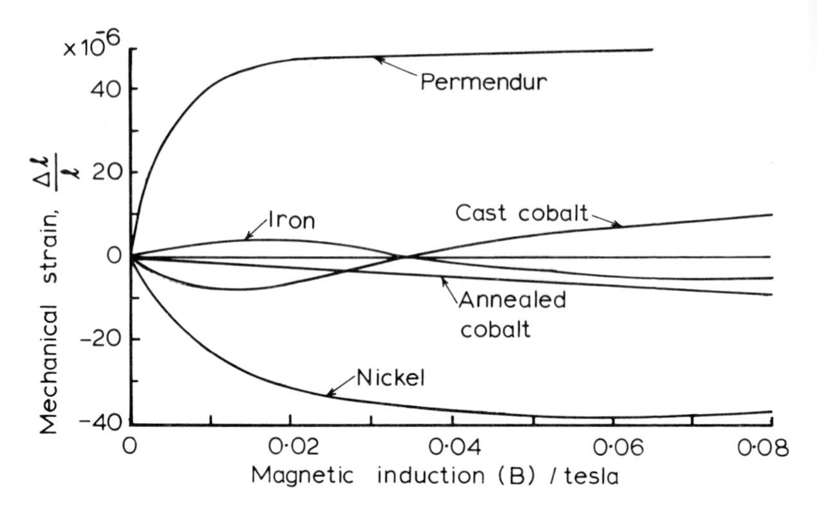

Fig. 4.11. Relationship between mechanical strain and applied magnetic field due to magnetostriction (From B. Carlin, *Ultrasonics,* (McGraw-Hill, New York, 1960).)

if the magnetization of such a material is changed a corresponding mechanical deformation will develop. A rod of a ferromagnetic material, such as iron or nickel, lying within a solenoid changes in length when a magnetic field is produced by switching on the current. Whether this change is an increase or a decrease in length will depend on the material of which the rod is made (fig. 4.11); it will not depend on the sense of the applied magnetic field. Nickel, for example, decreases in length when the field is applied. If an oscillatory electric current of frequency ν is passed through the coil, the length of the rod will then decrease and return to its original length once every *half cycle* of the electrical oscillations, and the rod can be regarded as a generator of mechanical vibrations of frequency 2ν. Various modifications to this simple scheme are incorporated in practice. The relationship between the mechanical strain and the applied magnetic field is illustrated in fig. 4.11 for a number of different materials. The effect is quite large in nickel, which has proved to be a very satisfactory material for use in magnetostrictive transducers. Difficulties may be experienced with the placing of the exciting coil, and also with the eddy-current losses which limit the useful frequency range to below about 100 kHz. The latter could be avoided by using ferrites, since these materials are non-metallic; however, because of their poor mechanical properties, they are seldom used in transducers. Magnetostrictive transducers are principally used in generators in high-power applications at low frequencies.

The inverse of the magnetostrictive effect, that is the change in the magnetization of such a material resulting from a mechanical deformation, can be exploited in transducers used as detectors.

Electromagnetic transducers are really technical developments of the conventional loudspeakers used at audio frequencies. They can be used at relatively low ultrasonic frequencies for high-power generation and they have been used at higher frequencies for internal friction measurements.

Lastly, we come to the electrostatic transducer. If a constant electrostatic potential difference V_0 is applied between the plates of a parallel plate capacitor (condenser), positive and negative charges appear on the plates and there is an attractive force between the plates. If an alternating voltage of amplitude less than V_0 and with frequency ν is superposed on this constant voltage, the attractive force will have a sinusoidal modulation at frequency ν. If one of the plates is 'loosely' supported it will move backwards and forwards at this frequency and generate mechanical oscillations in the surrounding medium. The force between the plates is

$$F = \frac{\epsilon_0 \epsilon_r A V^2}{2d^2} \tag{4.5}$$

where A is the area of each plate, d is the separation between the plates, and V is the potential difference between the plates. From this one can see that F is proportional to the relative permittivity ϵ_r of the material between the plates. Consequently the larger the value of ϵ_r, for a given amplitude of the applied voltage oscillations, the larger will be the amplitude of the ultrasonic wave that is generated. An electrostatic transducer can also be used as a detector of ultrasound. If a wave of mechanical vibrations with frequency ν falls on one plate of the capacitor (condenser) and that plate is free to oscillate, then as it oscillates the capacitance will oscillate at the frequency ν and these oscillations can be detected and amplified with the appropriate electrical circuitry. In other words, we have the principle of the 'condenser microphone'.

4.4. *Acoustic matching*

In Section 2.3 we considered the reflection and transmission of ultrasonic waves and the idea of acoustic impedance. Let us recall in particular eqn. (2.44):

$$\left.\begin{aligned} a_r^2 &= \left(\frac{w_A - w_B}{w_A + w_B}\right)^2 a_i^2 \\ a_t^2 &= \left(\frac{2w_A}{w_A + w_B}\right)^2 a_i^2 \end{aligned}\right\} \tag{2.44}$$

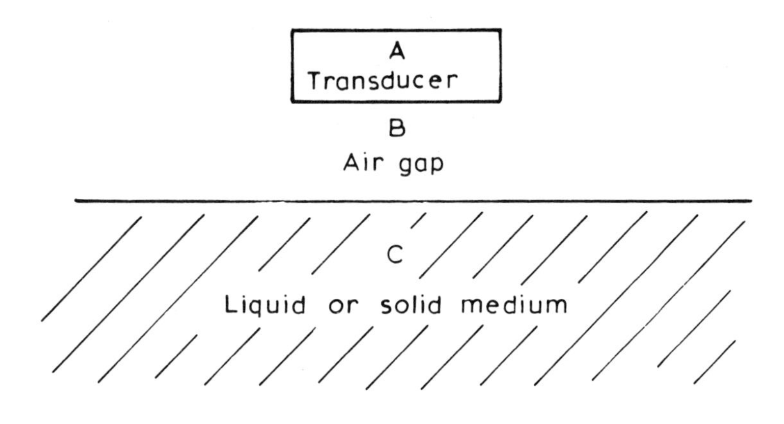

Fig. 4.12.

where a_i, a_r, and a_t are the incident, reflected, and transmitted amplitudes, respectively, for an ultrasonic wave incident in medium A on a boundary with medium B. We also noted that to obtain maximum transmitted energy, that is minimum reflected energy, requires $w_A = w_B$; the two media are then acoustically matched. The acoustic impedance of a gas is, typically, very much lower than that of a liquid or solid (see Table 2.1). Suppose that we try to use a transducer to propagate ultrasound into a liquid or solid medium and that there is an air gap between the transducer and the medium (fig. 4.12). There is a very bad acoustic mismatch between the transducer and the air gap and also a very bad acoustic mismatch between the air gap and the medium. Consequently very little ultrasonic power will be transmitted from the generator to the medium unless one establishes good acoustic contact between the transducer and the medium; the air gap should therefore be eliminated. For a liquid one simply immerses the transducer in it; for a solid good acoustic contact is made either by a film of liquid or grease between the transducer and the specimen or by cementing the transducer to the specimen using, say, epoxy resin (Araldite). This is for longitudinal waves. To propagate transverse waves into a solid the transducer must be cemented to the solid.

Suppose w_A and w_C are the acoustic impedances of a transducer and of the medium into which it is desired to launch ultrasonic waves, respectively. For maximum transmission to the medium, the thickness of the cement layer should be equal to one-quarter of an acoustic wavelength, and the acoustic impedance w of the cement should be equal to the geometric mean $\sqrt{(w_A w_C)}$ of the acoustic impedances of the transducer and the medium. The second condition comes from considering the product of the transmission coefficients

$$T = \left(\frac{2w}{w + w_C}\right)^2 \left(\frac{2w_A}{w_A + w}\right)^2 \qquad (4.6)$$

for the two boundaries and maximizing T as a function of w.

5. Pulse-echo techniques

5.1. *Asdic and sonar*

The possibility of the use of an ultrasonic echo method for the detection of submerged obstacles at sea, such as icebergs or wrecks, appears to have been suggested by L.F. Richardson at the time of the *Titanic* disaster in April 1912. This idea was pursued intensively during the 1914–18 War with a view to detecting enemy submarines, and a viable system was developed by P. Langevin. A diagram of Langevin's underwater ultrasonic generator is given in fig. 5.1; the same device was used as a detector. A beam of ultrasound is propagated vertically downwards in the sea, is reflected at the sea bed, or by some other object such as a submarine above the sea bed, and returns to the generator/detector (fig. 5.2). In the meantime the generator/detector has been disconnected from the electrical oscillator and has been connected to the detector circuit. By recording the departure of the original pulse and the arrival of the echo, using an oscilloscope, the transit time t for the return journey over a distance of $2d$ can be determined. Thus, assuming the speed c of ultrasound in sea water is known, the depth d can be calculated from

$$d = \tfrac{1}{2}ct. \tag{5.1}$$

Once a pulse has travelled to the sea bed and returned, the process can be repeated with another pulse of ultrasound. The interval between successive pulses must be greater than t, otherwise there will be an ambiguity as to which echo belongs to any given original pulse. Because the depths involved are quite large the repetition rate for successive ultrasonic pulses must be rather slow.

As used in the manner we have outlined simply for the measurement of a depth d, this device would be described as an ultrasonic depth-sounder. Such depth sounders are now widely used for measuring the depth of a river, lake, or sea in routine charting operations or as a navigational aid. If a submarine is situated above the sea bed with its motors switched off it will be producing very little sound and so is unlikely to be detectable by simply listening with a hydrophone (underwater microphone). However,

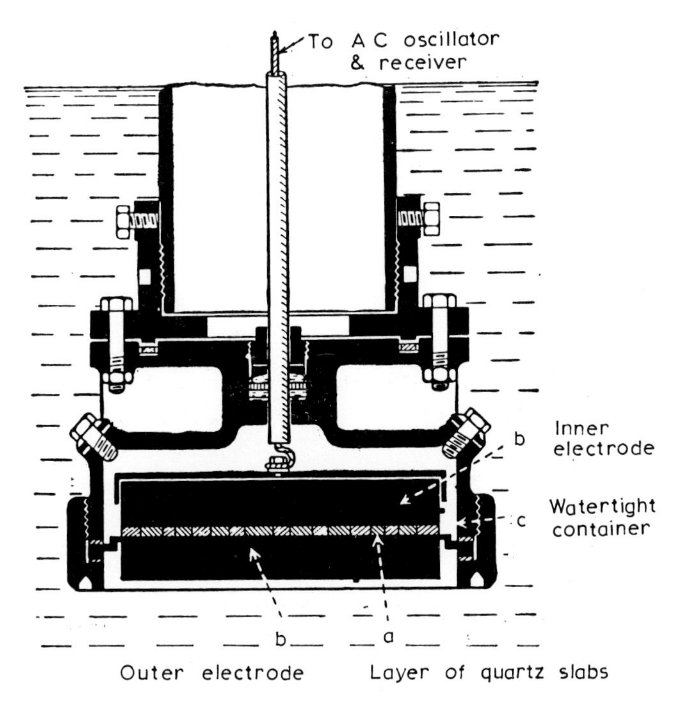

Fig. 5.1. Langevin's quartz ultrasonic transmitter and receiver.

the presence of the submarine could be detected by a sudden change in the value of *d*, measured by a depth sounder, and also by a change in the quality of the echo, as the ship with the ultrasonic equipment passes over the submarine. By the Second World War, the system originally pioneered by Langevin had been developed into standard naval equipment for submarine detection known as Asdic (from the initials of the *A*llied *S*ubmarine *D*etection *I*nvestigation *C*ommittee). In addition to this, ultrasonic listening devices were also mounted on torpedoes to provide guidance systems to enable them to 'home' on a submarine and destroy it.

Asdic constituted the first serious man-made application of ultrasound and the origins of its development, in the work of Langevin, preceded the development of radar by about 30 years. We have used the word 'man-made' because, as we shall describe later, ultrasound is actually used in an echo-location system by many different kinds of animals; but this was largely unsuspected in the days when Asdic was being developed. However, the development of radar, during the War of 1939–45, resulted in many improvements in electronic technology. These

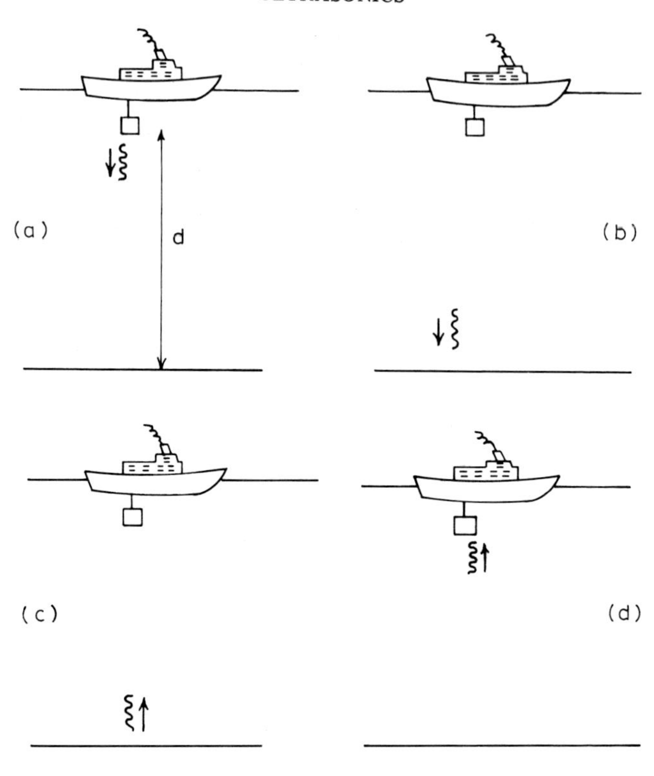

Fig. 5.2. Pulse-echo depth sounding.

developments in turn completely revolutionized the Asdic submarine detection system by introducing the possibility of using arrays of transducers, enabling sector-scanning to be introduced, and by improving visual display systems. The principle involved in scanning with a linear array of transducers is illustrated in fig. 5.3. The 21 transducers shown in fig. 5.3(a) generate pulses of ultrasound in succession, from left to right, with a delay of $(d \sin \theta)/c$ between adjacent transducers. A plane ultrasonic wave is then propagated in the direction θ (see fig. 5.3(b)); this is a rather nice illustration of Huyghens' principle. The angle θ is varied during a scan by changing the length of the time delay. The reflected ultrasonic pulses are detected with the same array of transducers. Ultrasonic underwater detection and ranging systems using sector scanning, with many other applications in addition to detecting submarines, then came to be known as *sonar* (*so*und *n*avigation *a*nd *r*anging) in imitation of radar.

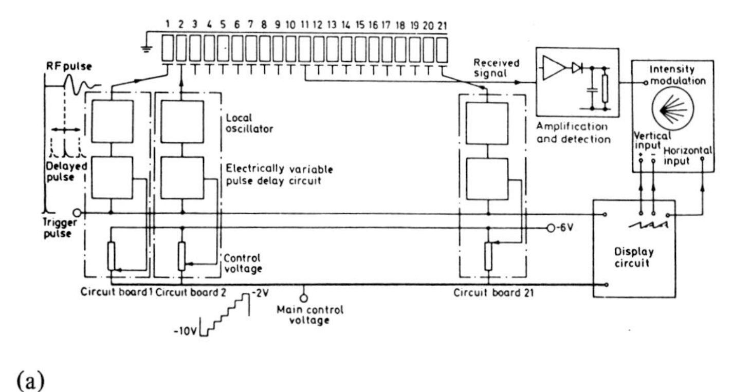

(a)

(b)

Fig. 5.3. (*a*) Block diagram of sector-scanning system and (*b*) diagram to illustrate the principle involved in the generation of a plane ultrasonic beam in a direction θ. (From J.C. Somer, 1968, *Ultrasonics,* **6**, 153.)

Radar, which was originally developed for the detection and location of aircraft, cannot be used for underwater work because of the very rapid attenuation of radio waves in water. On the other hand, as we have seen in Chapter 3, the attenuation of sound, or of ultrasound, is much lower in liquids than in gases. Consequently, although sonar can operate over very long distances under water, it can only operate over rather short distances in air. We have seen in Section 3.2 that in a liquid or gas the attenuation of ultrasound is greater at higher frequencies than it is at lower frequencies (the attenuation coefficients were quoted as proportional to ν^2). Therefore in marine applications, where very long path-lengths are involved, or in airborne applications, one uses low ultrasonic frequencies in order to avoid having too high an attenuation; frequencies of the order of tens of kHz are therefore used. If one is looking at large objects, such as the sea bed or a submarine, resolution is not likely to be a serious problem and so one can afford the poor resolution that goes with a low frequency, i.e. a long wavelength (see Section 7.3).

Although underwater ultrasonic systems were orginally developed for the detection of submarines, more peaceful underwater uses of ultrasonic ranging have been developed since then. This may involve an echo-sounder with a fixed vertical beam that is, essentially, used for measuring the depth of the sea or of a lake or river, as we have already mentioned. Alternatively, a scanning system may be used and the echo signals can then be processed to produce a visual display showing bearing and range like a radar display. In these applications, it is merely the fact that reflections occur which is important; the nature of the sea bed or of an obstacle is immaterial. However, by making the equipment more refined, it is possible either to study variations in the nature of the sea bed itself or to try to detect objects on or near the sea bed. Thus ultrasonic pulse-echo systems can be used to search for wrecks or for shoals of fish, and in the off-shore oil and gas industries for the inspection of installations and the monitoring of divers.

Although ultrasound can travel very great distances underwater it does, nevertheless, give a detectable echo from small objects such as fish. Several successful systems have been developed for the detection of fish and monitoring of trawls. One of these, the Humber system, which operates at 30 kHz, is illustrated in fig. 5.4. It uses an echo-sounder with a vertical beam and the angle of the beam is about 20°, which is sufficient to maintain good reception in spite of the rolling of the ship. The duration of each pulse is 500 μs and the peak electrical power output from the transmitter is 8 kW, which is extremely high for an echo sounder. The system is capable of detecting a single fish 35 cm long at a depth of 460 m. The transducer may be lowered beneath the hull of the ship; the purpose

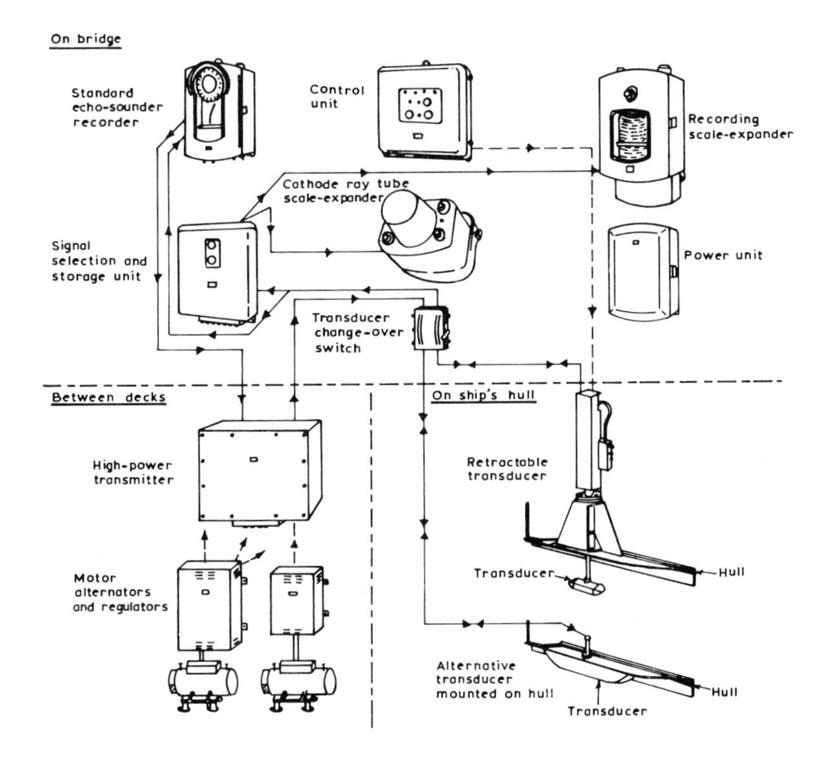

Fig. 5.4. Fish detection system (By G.H. Ellis, P.R. Hopkin, and R.W.G. Haslett, in Proceedings of the Second World Fishing Gear Congress, 1963, pp. 363–6, (Fishing News (Books) Ltd.).)

of this is to improve the acoustic coupling to the seawater by avoiding the layer of aeration around the hull. For inshore fishing, in shallower waters, other systems using nearly horizontal rotating or scanning beams are used (fig. 5.5). There has also been considerable study of the ultrasonic echoes from different kinds of fish, the object being to be able to identify the size and species of fish from their echoes. For this work it may be desirable to work at higher frequencies to improve the resolution.

Although it was originally developed for underwater detection and location work, the ultrasonic pulse-echo technique has subsequently been adapted to operate in other media, both gaseous and solid as well as liquid, and for an extremely wide variety of purposes. A recently introduced ultrasonic echo-ranging automatic focusing camera based on the pulse-echo technique is marketed by the Polaroid Corporation.

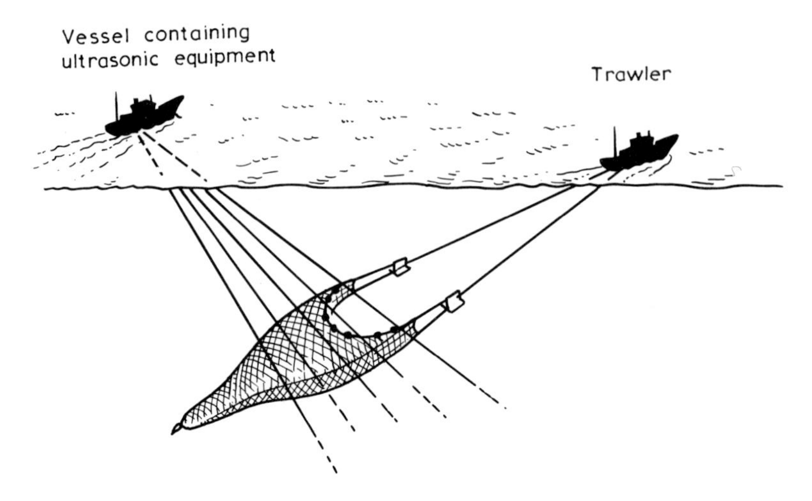

Fig. 5.5. Artist's impression of a midwater trawl in action under observation with bifocal equipment from abeam (from G.M. Voglis and J.C. Cook, 1966, *Ultrasonics,* **4**, 1).

5.2. *Thickness measurement and flaw detection*

In this section we shall be concerned with some applications of pulse-echo techniques in non-destructive testing. Whereas in marine applications (see Section 5.1) rather low ultrasonic frequencies are used because it is range rather than resolution which is important, in flaw detection it is resolution rather than range that is required and so much higher frequencies are used. For this work it is customary to use longitudinal ultrasound with frequencies in the region of 1 or 2 MHz and to use a single transducer, suitably mounted in a probe, acting as both generator and detector. Consider a specimen with two parallel plane surfaces with such a probe mounted on one surface (fig. 5.6(*a*)). Suppose that the probe is connected, via an amplifier, to the *y* plates of an oscilloscope and that the time-base of the oscilloscope is triggered at the moment when the probe launches a short ultrasonic pulse into the specimen. If the specimen contains no flaws (fig. 5.6(*a*)) the pattern shown in fig. 5.7(*a*) should appear on the screen; the peak marked A corresponds to the original pulse while C corresponds to the pulse reflected from the bottom of the specimen. The trace on the screen will, of course, die away and so, in order to obtain a steady trace, the pulse will have to be repeated at a rate greater than about 20 Hz. However, the repetition rate must not be too rapid because it is desirable that the echo (or echoes) from any given pulse should be received before the next pulse is emitted by the

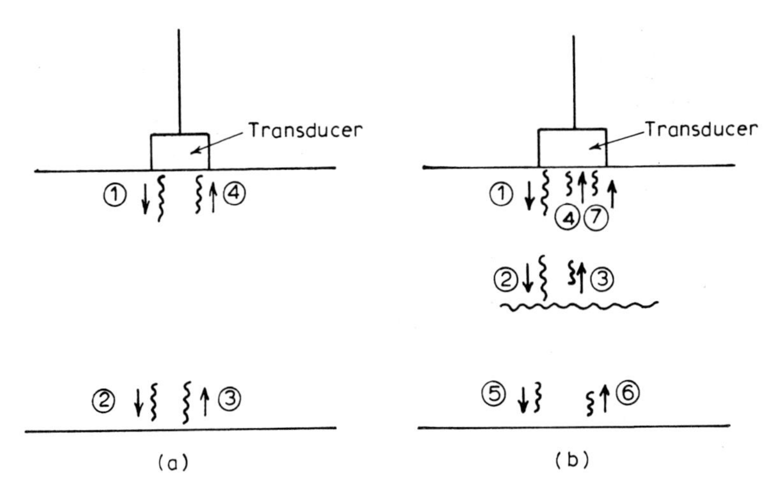

Fig. 5.6. Principle of pulse-echo method of flaw detection.

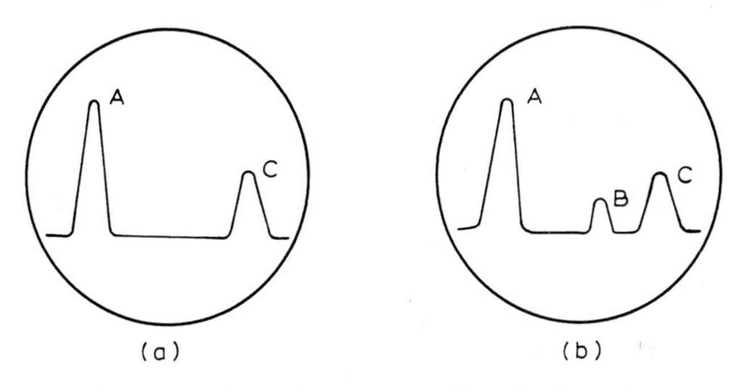

Fig. 5.7. Schematic representation of echo pattern.

transducer. The situation illustrated in fig. 5.7(*a*) is rather an over-simplification, for in practice the echo C will be followed by a train of further echoes corresponding to successive double transits of the specimen's thickness. The amplitude of these echoes will decay exponentially (fig. 5.8).

Assuming that the velocity in the material of the specimen in fig. 5.6(*a*) is known, the measurement of the time interval between A and C enables the thickness of the specimen to be calculated. At first sight this may not seem very useful if one imagines that the thickness could probably be determined more easily by using a ruler or calipers. However, in practice the ultrasonic pulse-echo method is now quite widely employed in thickness gauging in the continuous monitoring of the output from sheet-production processes for materials such as metals or plastics. One important

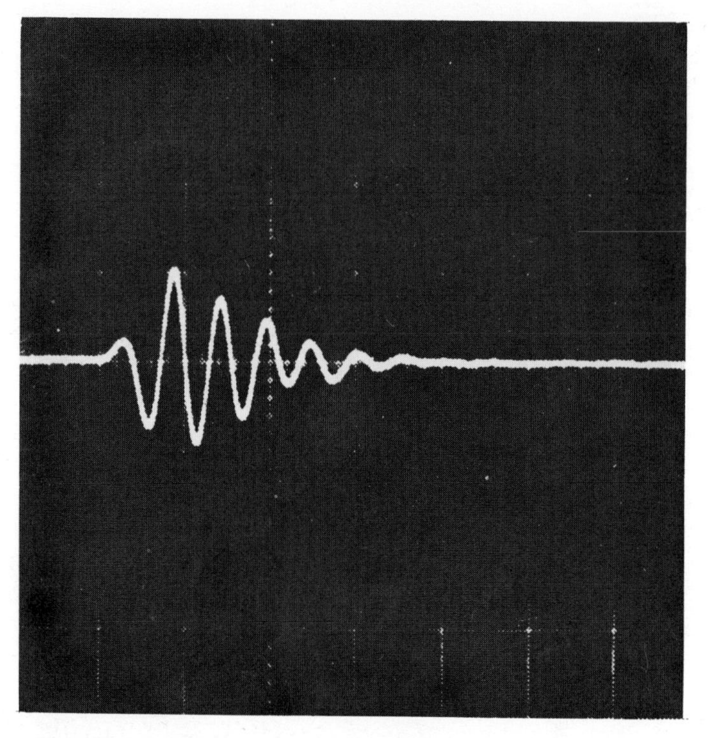

Fig. 5.8. An example of an echo train (from P.N.T. Wells, *Physical Principles of Ultrasonic Diagnosis* (Academic Press, London, 1969)).

reason for this is presumably because electronic control devices, which are needed if one wishes to make the process automatic, can then be coupled to the ultrasonic thickness gauge more easily than to other, more mechanical, types of thickness gauge. The pulse-echo technique can also be very useful when one surface of a specimen is inaccessible. Thus it is possible to monitor the thickness of pipe walls during manufacture or to determine the extent of corrosion in pipes in some system that it would be undesirable or inconvenient to have to dismantle for examination. It is also possible to study the extent of corrosion on the outside of the hull of a ship, from measurements made entirely within the ship.

Another application involving inaccessible surfaces, which has found widespread practical application, is in the determination of the thickness of the subcutaneous layer of fat in livestock, especially pigs and cattle. The boundary between the fat and the muscle (the lean meat) reflects ultrasound and this enables the thickness of the fat to be determined. The reflection from the boundary between the muscle and the bone enables the thickness of the lean meat to be determined. By moving the probe around over the surface of the animal one can scan a cross-section of the animal. We shall consider the use of scanning procedures in a little more detail when we come to consider the use of pulse-echo techniques in medical diagnosis in Section 5.3.

We suppose now that we use a specimen which contains a flaw (fig. 5.6(*b*)) in the path of the beam. This will give an additional reflected pulse B which will be recorded on the screen between pulses A and C (fig. 5.7(*b*)). The depth of the flaw can be estimated from the relative positions of A, B, and C on the screen. If the flaw is very much larger than the cross section of the ultrasonic beam, its size can be estimated by moving the probe around on the top surface of the specimen. The whole specimen can be examined for further flaws by moving the probe all over the top surface of the specimen and observing the echo pattern on the oscilloscope. Information about the shape and nature of a flaw can be obtained by varying the direction of the ultrasonic beam. In many industrial applications, however, one is concerned primarily, with the detection of flaws and obtaining rough estimates of their sizes, rather than with making very precise measurements of their locations. In certain circumstances, there may be some advantage in having separate generators and detectors, or in using shear waves or surface waves rather than longitudinal waves.

The pulse-echo method is widely used in the laboratory, workshop, and factory for the examination of specimens or artifacts for the purpose of detecting flaws. Rather than attempt to give an exhaustive account of all the current applications we shall just mention one particular area, namely the use of ultrasonic flaw-detection systems on railways, as an example.

The British Railways Board has used ultrasonic pulse-echo hand sets for many years in the routine testing of the running rails and of the wheels and axles of locomotives and rolling stock. Figure 5.9. shows railway axles (with deliberate flaws in them) being probed by trainee operators. Ultrasonic beams may be launched longitudinally through the end faces or obliquely across the shaft. More recently a special train has been developed to test the running rails while travelling at speeds of up to $40\,\text{km}\,\text{h}^{-1}$ (fig. 5.10). The scanning probes are mounted on a trolley beneath the car nearest to the camera. These probes slide along the surface of the running rails on a thin film of water which serves the dual purpose of providing lubrication and of maintaining good acoustic contact with the rails. A permanent record of the ultrasonic echos is produced; this record is subsequently analysed so that any appropriate repair or replacement work can be initiated. A recent further development

Fig. 5.9. Railway axles having deliberate flaws in them being probed during the training of operators. (Photograph by courtesy of British Railways Board).

Fig. 5.10. Ultrasonic test train. (Photograph by courtesy of British Railways Board).

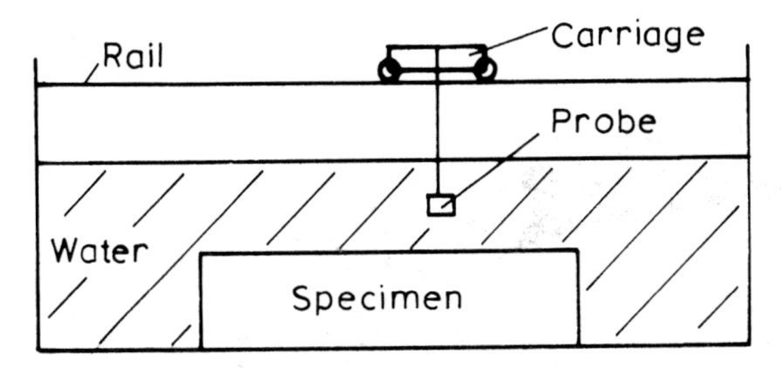

Fig. 5.11.

is to fit an ultrasonic system into a special rail assembly so that the wheels of a moving train can be tested.

Although the majority of the applications of the ultrasonic pulse-echo method for thickness measurement and flaw-detection are concerned with metallic specimens, numerous other materials can also be studied by this method. These materials include porcelain, glass, plastics, rubber, wood, leather, natural rock, synthetic stone, and concrete. The chief condition to be satisfied is that the material must be sufficiently transmissive for ultrasound with frequency between about 1 and 2 MHz, to enable useful echoes to be obtainable.

In both thickness measurement and in flaw-detection it is necessary to maintain good acoustic contact between the probe and the specimen (see Section 4.4). This is achieved by ensuring that the surface of the specimen is clean and by maintaining a film of water or oil between the probe and the specimen; in addition to providing good acoustic coupling, the film of liquid also acts as a lubricant. Alternatively, it may be more convenient to immerse the specimen and the probe in a bath of some liquid, usually water; in this case there is no need for the probe to be particularly close to the specimen, provided the distance between the two is kept constant while the specimen is being scanned (fig. 5.11).

5.3. *The pulse-echo method in medical diagnosis*

The differences between the acoustic impedances on opposite sides of many boundaries within the human body are large enough for quite significant reflection of ultrasound to occur at these boundaries (see Table 5.1). Ultrasound is particularly useful for the examination of internal parts of the body for which it would be inappropriate or undesirable to use X-rays. By 'inappropriate' we refer to the examination of soft tissues because X-rays are rather insensitive to small differences in the density of soft tissues. By 'undesirable' we refer to those parts of the body that are particularly susceptible to permanent damage by X-rays.

Table 5.1. Reflection of longitudinal waves incident normally on the plane boundary between two media, expressed in decibels below the level from a perfect reflector.

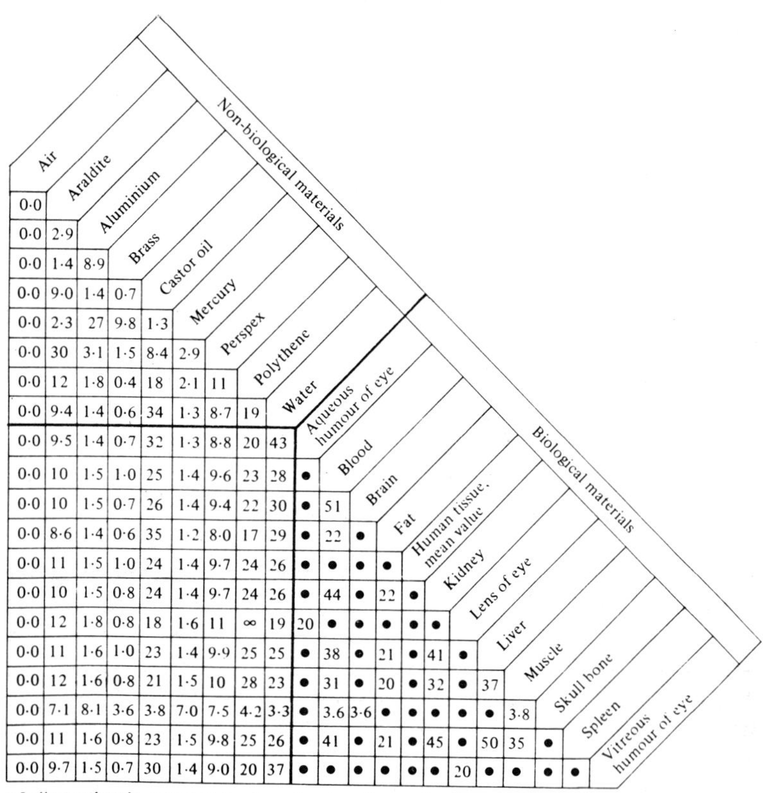

Non-biological materials: Air, Araldite, Aluminium, Brass, Castor oil, Mercury, Perspex, Polythene, Water

Biological materials: Aqueous humour of eye, Blood, Brain, Fat, Human tissue, mean value, Kidney, Lens of eye, Liver, Muscle, Skull bone, Spleen, Vitreous humour of eye

	Air	Araldite	Aluminium	Brass	Castor oil	Mercury	Perspex	Polythene	Water	Aqueous humour of eye	Blood	Brain	Fat	Human tissue, mean value	Kidney	Lens of eye	Liver	Muscle	Skull bone	Spleen
Araldite	0·0																			
Aluminium	0·0	2·9																		
Brass	0·0	1·4	8·9																	
Castor oil	0·0	9·0	1·4	0·7																
Mercury	0·0	2·3	27	9·8	1·3															
Perspex	0·0	30	3·1	1·5	8·4	2·9														
Polythene	0·0	12	1·8	0·4	18	2·1	11													
Water	0·0	9·4	1·4	0·6	34	1·3	8·7	19												
Aqueous humour of eye	0·0	9·5	1·4	0·7	32	1·3	8·8	20	43											
Blood	0·0	10	1·5	1·0	25	1·4	9·6	23	28	•										
Brain	0·0	10	1·5	0·7	26	1·4	9·4	22	30	•	51									
Fat	0·0	8·6	1·4	0·6	35	1·2	8·0	17	29	•	22	•								
Human tissue, mean value	0·0	11	1·5	1·0	24	1·4	9·7	24	26	•	•	•	•							
Kidney	0·0	10	1·5	0·8	24	1·4	9·7	24	26	•	44	•	22	•						
Lens of eye	0·0	12	1·8	0·8	18	1·6	11	∞	19	20	•	•	•	•	•					
Liver	0·0	11	1·6	1·0	23	1·4	9·9	25	25	•	38	•	21	•	41	•				
Muscle	0·0	12	1·6	0·8	21	1·5	10	28	23	•	31	•	20	•	32	•	37			
Skull bone	0·0	7·1	8·1	3·6	3·8	7·0	7·5	4·2	3·3	•	3·6	3·6	•	•	•	•	•	3·8		
Spleen	0·0	11	1·6	0·8	23	1·5	9·8	25	26	•	41	•	21	•	45	•	50	35	•	
Vitreous humour of eye	0·0	9·7	1·5	0·7	30	1·4	9·0	20	37	•	•	•	•	•	•	20	•	•	•	•

• Indicates that the corresponding boundary is unlikely to be of practical interest

(From P.N.T. Wells, *Physical Principles of Ultrasonic Diagnosis* (Academic Press, London 1969)).

Ultrasound is, therefore, now being used quite extensively for the examination of organs such as the brain, eyes, breast, heart, and liver and also in obstetrics. The advantage of X-rays, where they can be used, is the directness of the manner in which they give rise to a permanent record on a photographic film. Ultrasonic pulse-echo investigations, however, have to be performed by a scanning procedure leading to the production of a picture on an oscilloscope; this then has to be photographed if one requires a permanent record.

There are several different ultrasonic scanning procedures that can be used to build up a detailed picture of an otherwise inaccessible object, discontinuity, or boundary—particularly, but not exclusively, with applications to medical examinations. The terms 'A-scan', 'B-scan', and, occasionally, 'C-scan' are used in the literature.

In the A-scan the amplified echo signal is applied to the y plates of an oscilloscope while the time base is applied to the x plates, see fig. 5.12(*b*). This is very similar to the procedure frequently used in thickness measurement and flaw-detection (see fig. 5.7). The positions of the various reflected pulses on the screen can be used to determine the depths of the various boundaries below the surface. There are many different diagnostic applications of the A-scan. We shall just describe two examples, namely the localization of brain mid-line structures, and the measurement of the axial length of the eye.

In a normal healthy person the mid-line structures of the brain lie in the plane which is the geometrical plane of symmetry of the skull. If an A-scan is performed with an ultrasonic probe placed in contact with the skull of a patient, just above one of the ears, the ultrasonic reflection from the mid-line structures can be identified because it is normally the largest echo from within the skull. In a normal patient this echo will be at the centre of the A-scan. But in a patient suffering from an (intra) cerebral haemorrhage the mid-line is likely to be displaced to one side. This displacement may be detected because the ultrasonic echo from the mid-line structures will then not be in the centre of the A-scan. Conclusions based on a single A-scan are not very reliable and it is much more satisfactory to examine the patient's head from both sides in turn. These two scans can be done in rapid succession by using two probes, one on either side of the head. An improved technique involves using these two probes to locate the mid-line structures both by reflection and by transmission. This is illustrated in fig. 5.13 in which four scans are displayed simultaneously; each of the two probes is used separately for reflection measurements and the two probes are used together for transmission measurements in both directions (fig. 5.13(*a*)). Figures 5.13 (*b*) and (*c*) show results obtained for a normal patient and for a patient with a displaced mid-line structure.

For measuring the axial length of the eye by a pulse-echo technique, the difficulties involved in establishing good acoustic contact have led to the development of special probes, such as the water delay probe illustrated in fig. 5.14. In both of these examples, as in all other ultrasonic

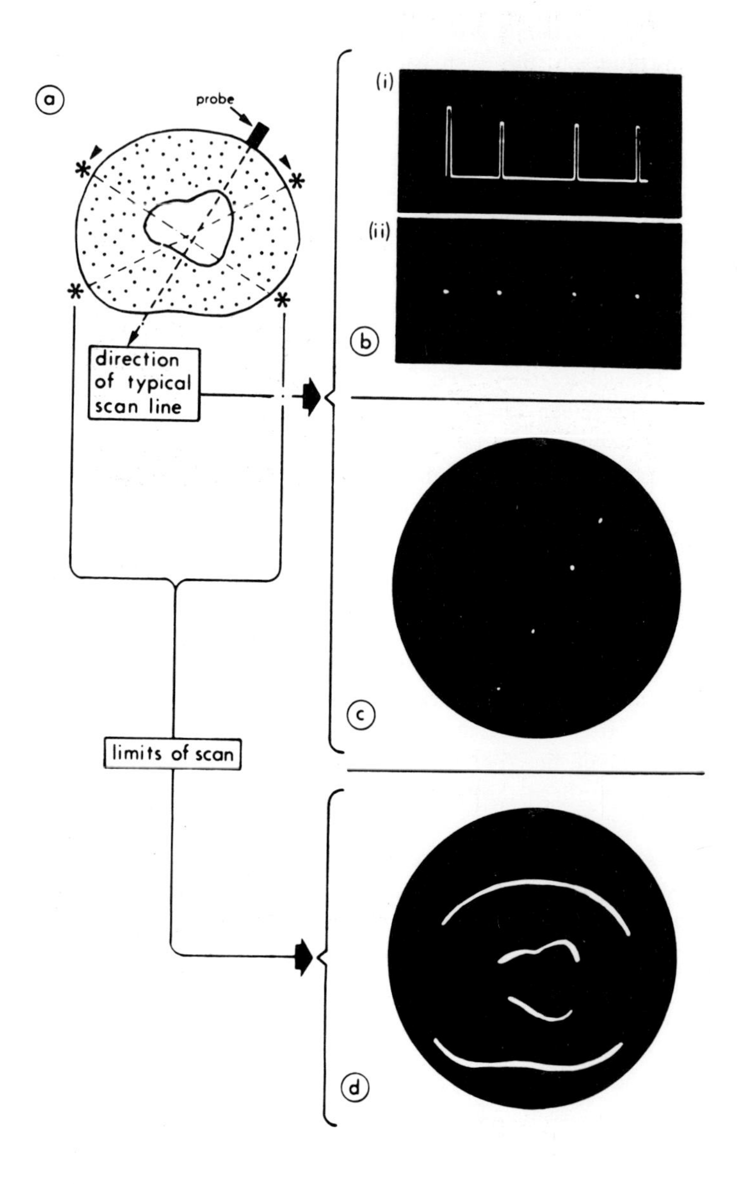

Fig. 5.12. Illustration of A-scan and B-scan. (*a*) schematic horizontal cross section through a human trunk, (*b*) (i) A-scan and (ii) B-scan for a typical horizontal direction, (*c*) B-scan of same direction as in (*b*), and (*d*) compound B-scan (From P.N.T. Wells, *Physical Principles of Ultrasonic Diagnosis* (Academic Press, London, 1969)).

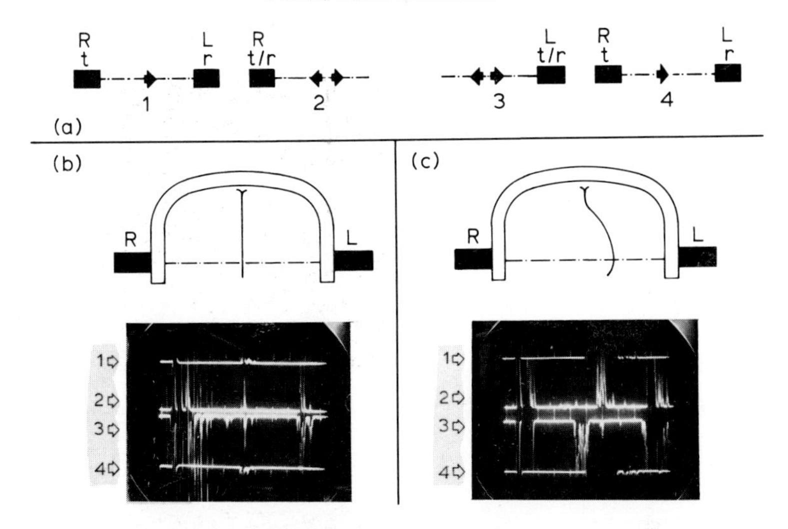

Fig. 5.13. A-scan localization of brain mid-line structures. (*a*) diagrams indicating positions of probes in standardized examination (L = left, R = right, t = transmitting probe, r = receiving probe). (*b*) scans for patient with central brain mid-line structures, (*c*) scans for patient with displaced brain mid-line structures. The time markers on the scans correspond to 10 mm distances in soft tissues. (Photographs by courtesy of P.N.T. Wells.)

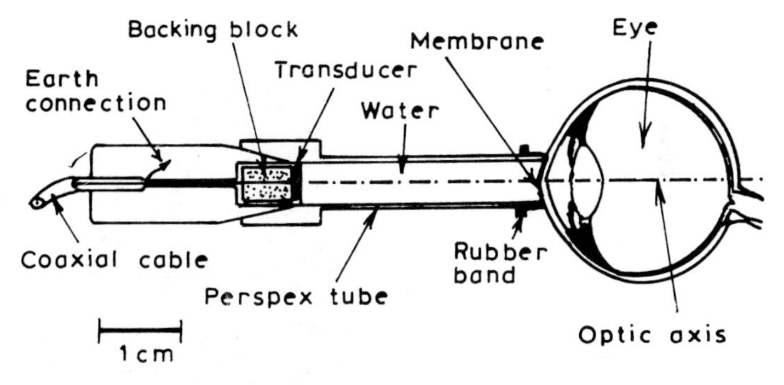

Fig. 5.14. Water delay ultrasonic probe for use in measuring the axial length of the eye. (Adapted from G.A. Leary, 1967, *Ultrasonics,* 5, 84.)

diagnostic applications, it is necessary to establish good acoustical contact between the probe and the patient. This is done either by using a thin film of liquid between the probe and the patient, as in the encephalography example, or by using a water bath, as in the examination of the eye.

In the B-scan the signal from the reflected ultrasonic pulse is applied

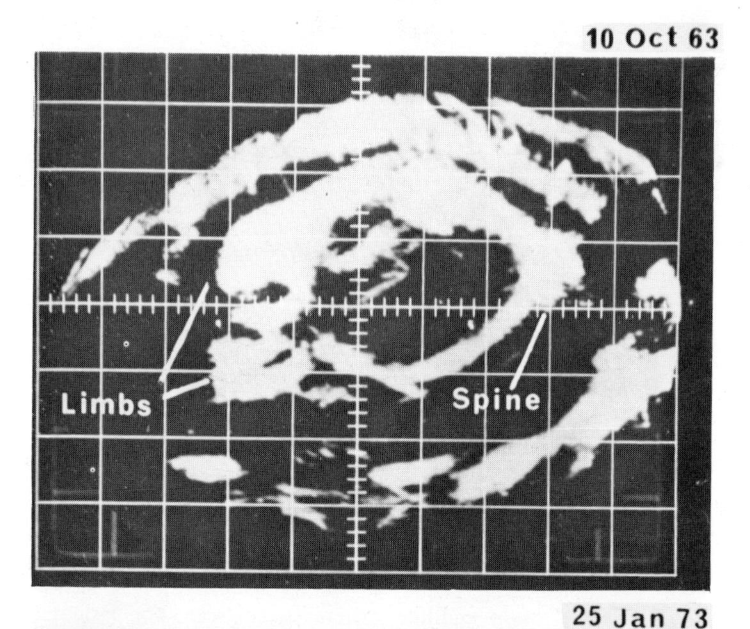

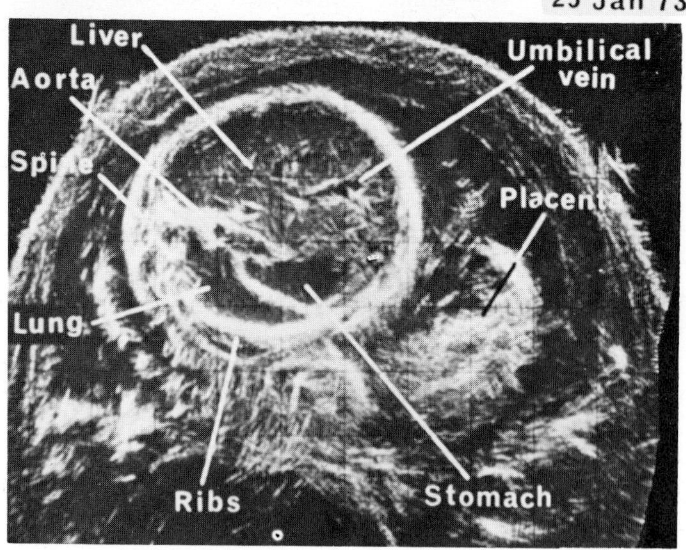

Fig. 5.15. Ultrasonic pulse-echo B-scans of a foetal trunk obtained in (*a*) 1963 and (*b*) 1973. (From G. Kossoff, *Ultrasonics International 1973 Conference Proceedings* p. 173 (IPC Science and Technology Press, London, 1973)).

to the cathode of the oscilloscope, thereby modifying the intensity of the spot. Figure 5.12 (*b*) illustrates schematically the distinction between a simple A-scan and a simple B-scan. By using a compound B-scan, integrated from many individual scans, one can reconstruct a pictorial representation of a cross section of the object under examination; fig. 5.12(*d*) shows a schematic output from a compound B-scan of a horizontal cross section of the trunk of a human body. Thus the compound B-scan can produce a much closer analogy to an X-ray (transmission) photograph than can be produced by an A-scan, so that a B-scan may be much easier to interpret than an A-scan. The numerous examples of the use of B-scans in medical diagnostic work include the examination of the breast, rectum, liver, heart, eye, uterus and brain. Perhaps the best-known of these is in obstetrics and gynaecology. With ultrasonic B-scans, it is possible to detect twins at a much earlier stage of pregnancy than would otherwise be possible, to detect abnormalities in the foetus, to determine the position of the placenta, and in the diagnosis of hydatidiform moles, ovarian cysts, fibroids, and many other gynaecological conditions. The examples which are reproduced in fig. 5.15 show how the sensitivity of the technique was improved over a period of about ten years.

In the C-scan both the x and y plates of the oscilloscope are used to refer to coordinates of the positions of the transducer and the signal from the reflected pulse is again applied to the cathode to vary the intensity of the spot; the transit-time information is not displayed. This kind of scan is much less important than the A-scan and B-scan.

In all that we have said about the pulse-echo method it has been assumed that the system under examination is stationary. In the case of flaw-detection the flaws were assumed to be located in fixed positions in solid objects. In the medical diagnostic applications discussed in this section any movements of the parts of the body under examination would be considered a nuisance in that it would reduce the resolution of the technique. However, if it is desired to monitor the motions of a moving part of the body this can often be done by modifying the pulse-echo method to give a plot of position as a function of time. This can be done with an A-scan or a B-scan and a suitable pictorial display of the results can be obtained. The moving parts that can conveniently be examined in this way are principally in the heart or the brain. Also, moving parts can be studied using the Doppler effect from a moving reflector of ultrasound (see Section 2.2).

6. Ultrasound in Nature

It has been known for a long time that bats can 'fly in the dark', that is, not just that they can fly in the dark but that they can apparently find their way in the dark without colliding with obstacles in their paths. Moreover, whereas owls which also fly in the dark have very large eyes and can therefore reasonably be supposed to be using these to gather what little visible light may be available at night, the eyes of bats are not particularly large. By now it is probably quite widely known that bats use ultrasound to avoid obstacles when flying, by using what is effectively the pulse-echo method which we have described in the previous chapter. The generators and detectors of the ultrasound are quite different from the electronic generators and detectors that are used in most man-made applications. What is, perhaps, less widely known is that the use of ultrasound among animals is not restricted to bats nor is it restricted to navigational purposes. We shall outline both the early work on bats as well as more recent work on bats and on other animals.

6.1. *Early work on bats*

At the time that ultrasonic echo-location was 'invented' for the detection of submarines, nobody knew that the use of ultrasonic echo-location is actually quite widespread in Nature. There was, however, some indirect evidence for this, although much of it was old and almost forgotten. Lazaro Spallanzani, who lived from 1729 to 1799, performed experiments on the problem of obstacle avoidance by bats in the dark. He found that complete darkness, and even blindness, made no difference to the bats' ability to avoid obstacles. We quote from a translation of one of his letters which was written in 1793:

> We can blind a bat in two ways—either by touching and burning the cornea with a thin, red-hot wire, or by pulling the eyeball out and cutting it off. Sometimes the animal, suffering from the operation, flies with great difficulty; later it can be made to fly freely in a closed room either during the day or at night. During such flight, we observe furthermore that before arriving at the opposite wall, the bat turns and flies back dexterously avoiding obstacles such as walls, a pole

> set up across his path, the ceiling, the people in the room, and whatever other bodies may have been placed about in an effort to embarrass him. In short, he shows himself just as clever and expert in his movements in the air as a bat possessing its eyes. Only occasionally, because of fatigue, he tries to light on either the ceiling or the walls, and, if these are not too slippery, he can light there just as the seeing ones do.

In 1793 Spallanzani wrote letters to a number of other scientists, describing some of his experiments and asking other people to repeat them to check his results. One of these letters was read to the Geneva Natural History Society and one of the members of that society 'Monsieur Charles Jurine, a surgeon and great insectologist, ornithologist, and botanist' repeated some of Spallanzani's experiments to demonstrate them to the society. Jurine added something very significant of his own too; this was to demonstrate that if the *ears* of a bat were blocked with wax or some other material its power of avoidance of obstacles was very seriously impeded. Spallanzani repeated and confirmed this observation of Jurine's; Spallanzani's final conclusions on the matter have been summarized by Galambos:*

> The ear of the bat serves more efficiently for seeing, or at least for measuring distances, than do its eyes, for a blinded animal hurtles against all obstacles only when its ears are covered. . . . The experiments of M. Professor Jurine, confirming by many examples those which I have done, and varied in many ways, establish without doubt the influence of the ear in the flight of blinded bats. Can it then be said that . . . their ears rather than their eyes serve to direct them in flight? . . . I say only that deaf bats fly badly and hurtle against obstacles in the dark and in the light, that blinded bats avoid obstacles in either light or dark.

There was clearly a difficulty here, in that the flight of bats is apparently silent and yet their ears seemed to be so essential for the avoidance of obstacles. Spallanzani had considered the possibility that a tactile sense enabled bats to perceive obstacles, but he performed various experiments which seemed to eliminate this possibility. However, the idea that bats could use their ears to detect obstacles seemed completely inexplicable and most zoologists throughout the nineteenth century preferred to believe in some kind of "tactile" hypothesis.

For over a century following Spallanzani's death (in 1799) there was little advance in the understanding of the obstacle-avoidance of bats flying in the dark. The results of the experiments on blocking the ears of bats were forgotten and these results had to be rediscovered independently near the beginning of the twentieth century, apparently in complete ignorance of the earlier work of Spallanzani and Jurine. As Griffin†

*R. Galambos, 1942, *Isis*, **34**, 132.
†D.R. Griffin, *Listening in the Dark* (Yale University Press, 1958).

says, 'What was clearly missing in the thinking of all those concerned with these problems, up to and including Hahn,* was any clearly formulated realization that there could exist sounds inaudible to human ears but yet useful to a bat for obstacle avoidance.' The first clear appreciation of this possibility seems to be due to Hiram Maxim. Following the sinking of the *Titanic* in 1912 (see also Chapter 5) Maxim had suggested that echoes of sound of very *low* frequency (infrasound) could be used underwater by ships to detect and avoid icebergs and other ships; he suggested using a frequency of 15 Hz. Maxim's idea was prompted by 'the bat problem'. He seems to have been the first person to realize and to propound clearly that it was 'sound' outside the audible range that is used by bats in their 'blind navigation'. His reason for suggesting a low frequency was because that would correspond to the frequency of the wingbeats which he supposed to be the cause of the 'navigational' sound.

In the event, of course, it was eventually discovered that it is very high frequency sound (ultrasound) and not very low frequency sound (infrasound) which is caused by bats and which came to be used in underwater echo-location. However, low frequencies are now being used in geophysical work based on acoustic holography, see Section 7.5. The suggestion that it is high-frequency 'sound', or ultrasound, which is utilized by bats was made by Hartridge in 1920. Hartridge does not seem to have performed any experiments; also it is obscure as to whether he meant only high audible frequencies of about 15–20 kHz or genuine ultrasonic frequencies greater than 20 kHz. But he does seem to have been the first to appreciate the importance of having short wavelengths, i.e. high frequencies, in the perception and avoidance of small objects (see Section 7.3 on resolution).

6.2. *Bats and insects*

The first detection of ultrasound emitted by any animals seems to have been made by Pierce who discovered, in the 1930s, that some bush crickets emit ultrasounds. It was Pierce who had invented the 'Pierce oscillator circuit' (see page 67) and who was one of the pioneers of the instrumentation of ultrasonic generation and detection. However, what is probably more well known than this early experimental work on insects is the early experiments on the use of ultrasound by bats. In 1938 Pierce and Griffin demonstrated experimentally the emission of ultrasound by bats in flight. Subsequent work in the early 1940s indeed confirmed Hartridge's earlier suggestion that for their navigation bats depend on the emission of ultrasonic pulses and the interpretation of the echoes of these pulses

*About 1908.

received from objects forming potential navigational hazards. It was also demonstrated that the use of ultrasound by bats is not confined to the avoidance of obstacles in flight but is also used in hunting flying insects for food. The frequencies used are in the range from 30 kHz to about 100 kHz. A fascinating account of the earlier work on bats and their use of sonar, or echo-location, will be found in Griffin's book *Listening in the Dark*, quoted above.

There are about 800 different species of bats, grouped into 17 families; among the mammals this diversity is second only to that of the rodents. Thus in talking about bats one is concerned with a large assemblage of species which can be expected to have a variety of navigational requirements. It should, therefore, not be too surprising that there is no unique pattern to the echo-location behaviour of bats. Indeed some species of bats do not have any echo-location facility at all. Moreover, many different species which do practise echo-location use different patterns of ultrasonic pulse. Having established clearly the general features of bats' echo-location systems, people then turned their attention to determining the structures of the ultrasonic pulses used by different species of bats. We do not propose to describe these in great detail, but will make a few general points. Briefly, there are three main types of signal: short clicks, frequency-sweep pulses, and constant-frequency pulses.

Bats are divided into two sub-orders, Megachiroptera and Microchiroptera. The short clicks are found in the Megachiroptera while the frequency-sweep pulses and constant-frequency pulses are found in the Microchiroptera. The sub-order Megachiroptera consists of a single family of bats with about 130 species. They are distributed throughout the warmer parts of the Old World. Most of these species are 'fruit bats', but many of them eat flowers and pollen, while some of them are specialized for drinking nectar. Most fly visually, even in very dim light, without any form of echo-location. But if they are placed in *total* darkness, or if they are blinded, they cannot find their way. Of the Megachiroptera, only the genus *Rousettus* has been found to use echo-location in flying. The pulses produced are pairs of clicks. The duration of each click is of the order of 1–2 ms, the separation between the clicks in a pair is 20–44 ms and the number of pairs of clicks per second varies from about 2 to about 7. These clicks are audible because the frequencies present range from about 10 kHz to about 50 or 60 kHz. The sub-order Microchiroptera is divided into four super-families and 16 families and contains several hundred species. All these bats use echo-location and the frequencies involved are inaudible, i.e. ultrasonic. It would be naïve to suppose that so many different species of bats all use the same kind of ultrasonic pulses for echo-location and indeed they do not. Many species of bats use frequency-sweep pulses; for example a typical pulse from *Myotis lucifugus* started at 78 kHz and swept down to 39 kHz in 2·3 ms, having its maximum amplitude at a frequency of about 50 kHz. The advantage of a swept pulse, whether in radar or in sonar, is that it gives a very much better range resolution than a fixed-frequency pulse of the same duration, see Section 7.3. If the bat is 'cruising' it emits of the order of ten pulses per second, but in the final stages of manoeuvre the repetition rate may rise to as much as 150-200 pulses per second. In other bats the details of the frequency-sweep pulses may be different.

The constant-frequency pulse is used for a completely different purpose and is based on a different physical property, or effect, from that involved in the operation of sonar; namely, the Doppler effect. The sonar, or echo-location, which was discussed in Chapter 5 and in the earlier parts of this chapter, has been concerned with identifying the position of an object; this is done by measuring the delay between transmission of the original pulse and reception of the reflected pulse. The use of the Doppler effect enables a bat to determine the velocity of a moving object, relative to itself. If there is no relative motion of the source and the observer, and one is observing reflected sound or ultrasound that was generated with frequency ν, then the change in frequency, $\Delta\nu$, is given by

$$\Delta\nu = -\frac{2v_r}{c}\nu \qquad (2.36)$$

where v_r is the relative velocity of the reflector and the source (or observer). Therefore, by using a single steady transmitted frequency and comparing this with the frequency of the reflected ultrasound, one can determine the velocity of a moving object. The use of the (radar) Doppler shift is common in a number of man-made situations, such as radar 'speed traps' for motorists, and the ultrasonic Doppler effect is also exploited in one or two applications as well. There are many bat species which use constant-frequency pulses and which have been clearly shown to be using the Doppler effect.

The first example of the use of the Doppler effect by bats was detected in the Rhinolophidae, or horseshoe bats, by Möhres in 1953. The pulses emitted by these bats are very pure pulses of almost a single frequency, although there are small components at $\frac{1}{2}$ and $\frac{3}{2}$ of this frequency, indicating that the pulse is actually the second harmonic. Figure 6.1(*a*) illustrates an analysis of the pulse emitted by a cruising *Rhinolophus ferrumequinum*, while fig. 6.1(*b*) illustrates a rapid sequence of short pulses emitted when the bat is negotiating an obstacle or when it is landing. Evidence is now available to show that these bats are able to 'measure' Doppler shifts as small as a fraction of one per cent and are also able to

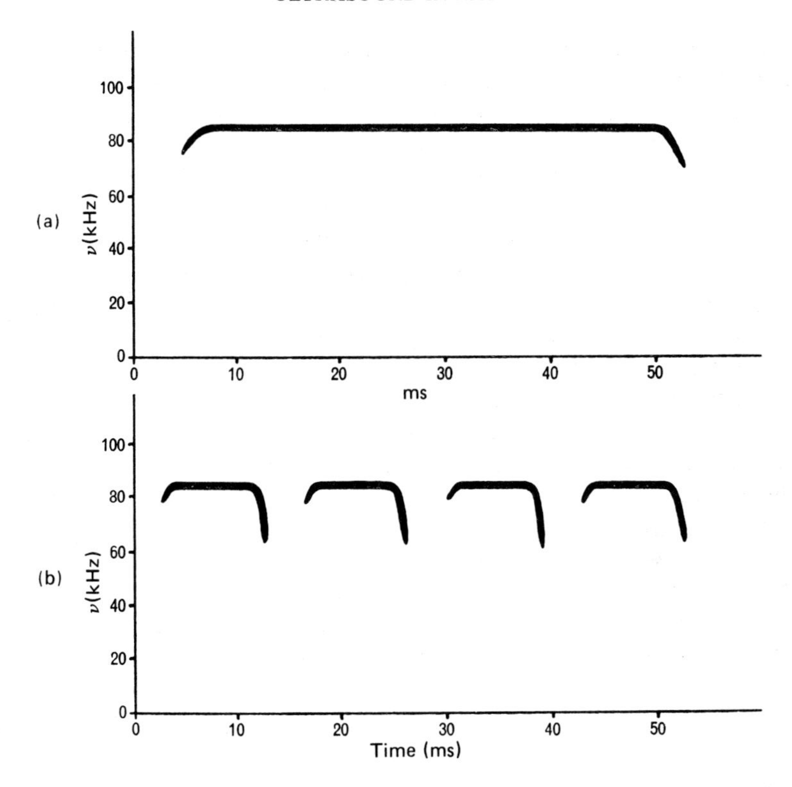

Fig. 6.1. Sonograms of 'constant frequency' pulses emitted by bat *Rhinolophus ferrumequinum*, (*a*) cruising pulse, and (*b*) part of a rapid train of shorter pulses. (From G. Sales and D. Pye, *Ultrasonic Communication by Animals*, (Chapman and Hall, London, 1974).)

control their emission frequency to the same order of accuracy. Other bats of the family Rhinolophidae exhibit similar ultrasonic behaviour, although different species use different frequencies, varying from about 40 kHz to about 120 kHz.

The division between bats that use frequency-sweep pulses and bats that use constant-frequency pulses is probably less sharp than the above account may seem to suggest. Thus some bats use pulses that are of constant frequency for part of the pulse and are frequency-swept for another part of the pulse. Moveover, in some species at least, bats seem to be able to alter their pulse structure to meet changes in their circumstances or needs.

In view of the comments made in Chapter 4 about the difficulties involved in making mechanical generators and, more especially, mechanical detectors for ultrasound, it is perhaps of interest to consider briefly how bats manage to generate and detect high-frequency ultrasound. (Clearly they do not use electrical oscillators, electronic circuits, and piezoelectric transducers.) Once again we distinguish between the Megachiroptera and the Microchiroptera. The pairs of short clicks, of rather low frequency, produced by the few of the Megachiroptera (*Rousettus*) that use echo-location are produced by the tongue and not by the larynx. This is rather unusual since it is the larynx which is used by most mammals in the production of sound. It is even rather uncommon among bats too, because the Microchiroptera use the larynx and not the tongue in the production of their ultrasonic pulses. In the larynx two lateral folds, the vocal cords, are brought together so that the respiratory passage is blocked. Then, as air pressure is built up, the cords part momentarily to allow a puff of air through and then they close again, the process being repeated cyclically. The frequency at which puffs of air are expelled between the cords depends on the length, mass, and tension of the vocal cords and constitutes the fundamental frequency of the voice, although the sound is modified by the mouth, lips, teeth, and nasal cavity and contains a rich sequence of harmonics. If we recall the expression for the fundamental frequency of vibration of a stretched string, namely

$$\nu = \frac{1}{2l}\sqrt{\left(\frac{T}{m}\right)} \tag{6.1}$$

where l is its length, T is its tension, and m is its mass per unit length, we see that a high frequency requires a light short string, maintained under very great tension. An examination of the larynx of a bat shows that it has evolved in such a manner as to meet these requirements, and it has now been demonstrated experimentally that it is the larynx that is involved in the production of ultrasonic pulses in the Microchiroptera. As to the detection of ultrasound by bats, all that we need to say here is that, as in the case of the larynx, so also in the case of the ear, this organ has evolved in bats so as to be capable of detection of high frequencies. The anatomical and other details of precisely how the ears of bats work are a subject of active study; these details differ considerably among the various species that use echo-location.

Let us return now to consider another of the experiments of Lazaro Spallanzani. Having shown that bats which had been blinded surgically could still navigate successfully (see page 92) he turned to the question of whether the eyes were important to bats in hunting for food. He captured 52 bats from a colony in the bell tower of the cathedral in Pavia (in Italy), blinded them and then released them. He returned four days later and captured another group of bats from the same colony. This was early in the morning when the bats had just returned with full stomachs after a night's hunting. Among the second group of bats he found four of the ones that had been blinded. He killed these bats, opened their stomachs, and found that they had swallowed just as many insects as the bats which had not been blinded surgically; in his own words '*La scoperta è bellissima*'—the discovery is very beautiful. The conventional interpretation that was put upon this result was to suppose that the bats locate and catch insects by listening to sounds produced by the insects. Although this may be part of the truth, it is by no means anything like the whole truth. Once it was established that bats use pulses of ultrasound in detecting and avoiding obstacles it was a reasonable extension to suggest that they might also use their ultrasonic pulses in hunting insects for food. This was indeed found to be the case; the experiments to demonstrate this involved taking cine-films synchronized with ultrasonic tape recordings. It appeared that the bats did not necessarily merely pursue an insect but were able to predict the insect's course and intercept it. It is also quite common for a bat to use its wings as a net or scoop to catch an insect and then transfer the captive to its mouth.

The use of echo-location by bats in the detection of nocturnal flying insects might suggest that hunting is a trivial matter for the bats. However, it was later discovered that many nocturnal insects are able to detect the bats' ultrasonic signals and take avoiding action. Observations of the behaviour of moths in the presence of bats showed that many moths try to evade an attacking bat by looping, spiralling or diving, but that other moths appear to take no evasive action at all and easily fall prey to an attacking bat. It has been demonstrated that many moths are able to detect artificial ultrasound over a wide range of frequencies, but especially in the range of 40–80 kHz. The fact that it is the ultrasonic pulses of the bats to which, at least, some moths respond can be demonstrated by using an artificial ultrasonic generator to produce pulses similar to those produced by bats. When the ultrasonic generator was switched on some moths began to respond by looping, spiralling or diving, just as in the presence of a hunting bat, but other moths ignored the ultrasound completely.

At the beginning of this section we mentioned the early discovery, by Pierce, that some bush crickets emit ultrasounds. For a long time the progress in the study of ultrasound produced by insects progressed very slowly by comparison with the work on bats. The bush crickets or tettigoniids (Tettigoniidae) produce sound in a completely different manner from that used by bats and marine mammals. They use a mechanism called *strigilation* which is illustrated in fig. 6.2. Sound, or ultrasound, is produced by a "file" and "plectrum" which, in the bush crickets, are

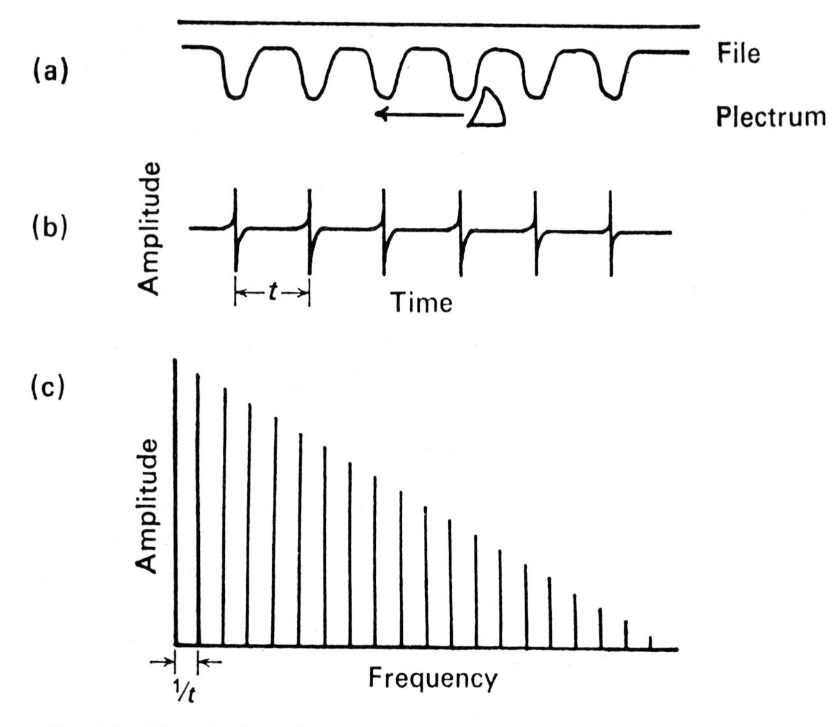

Fig. 6.2. The physics of strigilation, the interaction of the file and plectrum. (*a*) The plectrum moves along the file, striking each tooth in turn, (*b*) each tooth-strike produces a brief click, (*c*) a series of such clicks has a frequency spectrum with a fundamental at the tooth-strike rate and many harmonics. (From G. Sales and D. Pye, *Ultrasonic Communication by Animals*, (Chapman and Hall, London, 1974).)

both parts of the exoskeleton found near the base of the forewings. In insects, in general, any two parts of the body that can be brought together may be used for strigilation, and different insects use a wide variety of areas. As the plectrum moves across the file it produces a click each time it strikes a tooth. The frequency analysis of the sound produced contains a fundamental, which is the rate at which the plectrum strikes the teeth, together with a series of harmonics. The frequency spectrum that is produced is affected by a resonator coupled to the strigilation mechanism and much of the energy in the songs of the bush crickets is at ultrasonic rather than audible frequencies. The frequency range and the frequencies of maximum energy vary considerably among the different species of tettigoniids. It appears that the ultrasound they produce is not used for the purpose of echo-location. Rather, the songs of these insects, whatever the individual relative distribution of energy between the audible and the ultrasonic ranges for different species, are thought to serve principally as mating calls.

Insects produce sound in a variety of different ways. Dumortier* summarized the methods of sound production in insects as:

(*a*) friction of differentiated parts, as in tettigoniids which we have already mentioned;
(*b*) vibrations of membranes, for example the buckling of tymbal organs in cicadas and some moths;
(*c*) the expulsion of fluid (gas or liquid) as in the death's head hawk moth;
(*d*) shocks to the substrate, well known in the death watch beetle; and
(*e*) vibration of appendages, such as the wings, which produces the 'droning' sounds of bees and mosquitoes when in flight.

So far it is known that ultrasound can be produced by at least the first three of these methods. In addition to the tettigoniids, which we have already mentioned, several other groups of insects are also known to produce songs which, although they are clearly audible, also contain ultrasonic frequencies as well. These insects include crickets, grasshoppers, locusts, cicadas, and certain beetles and cockroaches.

6.3. *Birds, rodents, and marine mammals*

The ability of many species of bats to determine the distance, or the speed, or to some extent the nature, of an object by listening to the echoes of their own ultrasonic cries is the most well known, the most extensively studied, and the most clearly understood example of echo-location in nature. However echo-location is not confined to bats nor is it necessarily confined to the ultrasonic range. There is clear evidence of the use of echo-location by two different groups of birds, though using audible frequencies, and by certain dolphins which use a very wide range of frequencies in both the audible and the ultrasonic ranges. The birds are *Steatornis caripensis*, the oilbird of Caripe (in Venezuela) and certain swifts of the genus *Collocalia* which are the birds responsible for the production of the nests used by the Chinese in making birds' nest soup. The dolphins are the bottlenose dolphins, *Tursiops truncatus*. There is also some less definite evidence that several other species also indulge in echo-location.

*B. Dumortier, 'Morphology of sound emission apparatus in Anthropoda', in *Acoustic Behaviour of Animals*, edited by R.G. Busnel (chapter 11) (Elsevier, Amsterdam, 1963).

In addition to this, it has now been discovered that many different animals produce ultrasound, either on its own or in conjunction with audible sound, and that the purpose is not always echo-location. Although the songs of certain birds include ultrasonic components in addition to the audible sounds, it seems likely that birds are not able to hear at ultrasonic frequencies. Thus the oilbird of Caripe operates its echo-location system, or sonar, within the range of 6 kHz to 10 kHz. An interesting account of the establishment of echo-location by these birds will be found in chapter 11 of Griffin's book (see Further Reading). The *Collocalia* also use frequencies which are well within the audible range, although the actual frequencies vary from one species to another. Ultrasound has now also been detected in the calls of many rodents, in some cases on its own but in other cases in combination with audible sound. While there is some evidence of the use of ultrasound for echo-location purposes by certain rodents, it is principally used for general 'social' purposes; these include distress calls by the infants and in aggressive sexual and non-sexual encounters of adults.

Perhaps, after the bats, the most well-known example of echo-location is to be found among marine mammals. In view of the success which has attended the use of sonar in man-made marine applications it is, perhaps, not too surprising that underwater applications of the technique should occur in nature too. An extremely readable account of an important part of the work that was involved in studying the bottlenose dolphin, *Tursiops truncatus*, is given in the book by W.N. Kellogg (see opposite). The signals most commonly produced by this creature consist of successive series of rapidly repeated clicks, which contain audible frequencies as well as ultrasonic frequencies up to about 170 kHz. It was shown that these animals could navigate successfully to avoid obstacles in a pool, even when the possibility of visual detection of the obstacles was eliminated. The possibility of visual detection was eliminated by the turbidity of the water, or by performing the experiments at night, or by using transparent objects. Evidence was also obtained that these animals use their echo-location, or sonar, technique in hunting for food, not only to locate fish but also to discriminate between desirable and undesirable species of fish.

There are two other general features that should be mentioned in connection with the use of sonar in nature. The first concerns the possibility of 'jamming' the ultrasonic signals by using man-made ultrasonic signals. It is perhaps not too surprising that the echo-location of animals is not disturbed by the use of extraneous sources of ultrasound. What is perhaps more surprising is that when recordings of porpoises' echo-location pulses were played back while these animals were engaged in navigating an obstacle course these recordings seemed not to confuse the animals.

Similarly, it has been found that bats also exhibit a considerable degree of resistance to the jamming of their echo-location signals. However, if the extraneous ultrasound has sufficient intensity or a suitable frequency range, it is possible to cause some confusion to the bats and to reduce their echo-location efficiency. The second point is that one should not under-estimate the complexity and difficulty of the detection of the echoes. The intensity of an echo is necessarily very much lower than the intensity of the original pulse and it may be necessary to identify the echo in the presence of ultrasonic pulses received directly from other animals in the neighbourhood. Moreover, it is also necessary to distinguish between the echo from a given object and the echoes from other objects. As Kellogg* says

> Imagine, for example, the recording of a piano concerto that is being played on a hi-fi set. *Hearing* the recording is not much of a problem. Perceiving it as a concerto, identifying the composer, recognizing the particular work and movement, and even the recording artist, require more and more background information. Obviously, in the problem of perception, the intelligence and previous learning of the observer are of paramount importance. Echo-location is not just sensing the presence of an echo. It requires the ability to interpret, evaluate, and identify that echo. This complex avenue of auditory perception seems to be quite beyond the capacity of an ordinary man, who makes little use of it in his own surroundings.

6.4. *Blind guidance systems*

In Section 5.1 we have mentioned the development of sonar systems for marine navigational purposes and in the present chapter we have described the use of ultrasound by bats and other animals for obstacle avoidance and hunting for prey. This suggests that it might be possible to develop an ultrasonic system for use by blind persons as a navigational, or guidance, aid. In the first instance this may sound an attractive proposition because bats use ultrasound very successfully to 'see in the dark'. However, if one reflects on the descriptions of the pulse-echo method which we have given in Chapter 5, there is a serious difficulty in that the information obtained from ultrasonic scanning in a sonar, flaw-detection, or medical diagnostic system is usually converted electronically into a visual image. Such a visual image is obviously of little use to a blind person and so an ultrasonic guidance system for the blind has to convert the information received from ultrasonic echoes into some form that can be perceived by some sense other than sight. The commonest approach is to produce audible sound from the ultrasonic echoes.

In seeking to imitate the echo-location systems of bats etc. one should

*W.N. Kellogg, *Porpoises and Sonar*, (Chicago University Press, 1961).

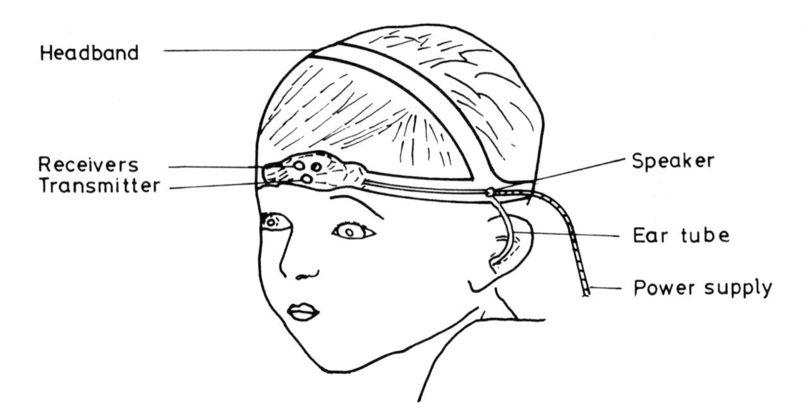

Fig. 6.3. Blind-guidance system which has been tested on a very small child. *Distance* is signalled by the *pitch* of the signal; low pitch means near, high pitch means distant. *Size* is signalled by the *amplitude* of the signal; low volume means small object, high volume means far away. *Texture* is signalled by the *clarity* of the signal; clear sound means hard object, fuzzy sound means soft object. *Right-left position* is given by *time-of-arrival* differences of a signal at the two ears. *Up-down position* is not given directly but can be ascertained by *head movements*. (Diagram by courtesy of T.G.R. Bower.)

recall that these echo-location systems have developed as the result of a highly specific evolutionary process. While it is no doubt possible to imitate these systems artificially, it is not enough just to imitate them. The artificial system also has to be sufficiently compact that it can easily be carried by a blind person. However, even to imitate the echo-location systems of bats, or other animals, may well not be enough to provide an adequate guidance system for a blind person because the blind person probably requires far more detailed information from a guidance system than does a bat.

For there to be any hope of success for an ultrasonic guidance system, the information obtained from ultrasonic echoes by conversion into audible signals must be of greater value than that which can be obtained by the blind person directly from audible signals generated by, or reflected from, obstacles. While several attempts have been made to develop ultrasonic blind-guidance systems and while they have achieved some degree of success, the fact remains that, for a number of reasons, no such ultrasonic system has yet found widespread adoption. One system which has been developed and tested using a very small blind child, less than one year old, is illustrated schematically in fig. 6.3. In this system the distance, size, and texture of an object are simulated by the pitch, amplitude, and clarity, respectively, of the signal at the ears; right-left position is indicated by the time difference between the signals at the two ears, while the up-down position is not given directly but can be ascertained from head movements. Of course, blind-guidance systems based on other carriers of information than ultrasonic waves have also been tried. However, the traditional use of a guide dog or the tapping of a stick are practices which have not yet been superseded.

7. Ultrasonic 'optics'

7.1. *Velocity and attenuation measurements*

When ultrasound is generated by the electrical excitation of a transducer we can take it that the frequency is usually known to a considerable degree of accuracy; if the ultrasound passes from one medium to another its frequency will not change. The other important parameters that one may need to determine are, first, either the wavelength or the velocity and, secondly, the attenuation coefficient. We divide the available experimental methods for determining these parameters into three groups:

(*a*) standing-wave methods
(*b*) pulse-echo methods
(*c*) other methods.

We shall discuss only (*a*) and (*b*) in this section; (*c*) includes a number of miscellaneous methods which are important in certain cases but are not of general interest.

(*a*) *Standing-wave methods*. Traditionally standing-wave methods have been used to determine the velocity of sound, and later of ultrasound, in gases. It is the wavelength which is actually measured and this is done by using a reflector to form standing waves and detecting the nodes or antinodes in the standing-wave pattern and measuring their separation. The velocity can then be found from the wavelength λ and the known frequency ν by using

$$c = \nu\lambda. \tag{1.5}$$

At audible frequencies the apparatus used for the measurements would be Kundt's tube (fig. 7.1). In Kundt's tube some convenient powder is sprinkled in the tube and when standing waves are excited the powder is agitated into little heaps at the nodes. By measuring the distance ($\frac{1}{2}\lambda$) between successive nodes the wavelength, and hence the velocity, can be determined. At audio-frequencies the wavelengths involved are such that the standing waves have a very suitable value of λ for this experiment, for example, in air $\lambda \simeq 66$ mm at a frequency of 5 kHz. However, at very much higher frequencies this method of detecting the nodes in the standing

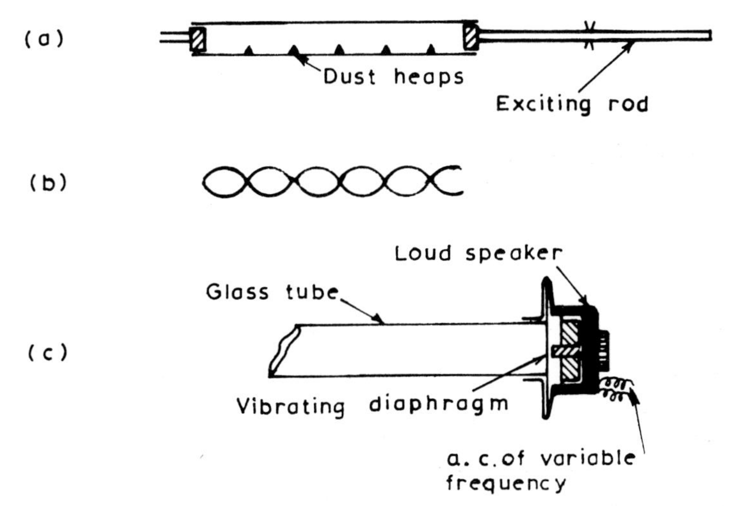

Fig. 7.1. Kundt's tube, (*a*) excited by stroking a rod fixed at its midpoint, (*b*) standing-wave pattern, and (*c*) excitation by a loudspeaker.

wave pattern becomes unsuitable because the wavelength becomes too small. Lord Rayleigh* mentioned experiments using sensitive flames which were capable of detecting ultrasonic standing wave patterns for wavelengths as small as 6 mm (i.e. frequency of about 55 kHz)†.

It is possible to form stationary ultrasonic waves in much the same way that stationary acoustic waves are formed in Kundt's tube. The stationary ultrasonic waves will be formed by having a transducer, which is excited electrically, as the source of ultrasound and by having a reflector to reflect the ultrasound. To form stationary waves the distance between the transducer and the reflector will have to be equal to an integral multiple of $\frac{1}{2}\lambda$ and so, to be able to adjust the distance to satisfy this condition, either the ultrasonic source or the reflector must be able to be moved in a controlled and calibrated manner. For measurements in air, or in other gases, one can use Pierce's interferometer which can be regarded as a development of Kundt's tube (fig. 7.2). The reflector can be moved backwards and forwards and is attached to a micrometer screw thread. As well as acting as a generator, the transducer can also be regarded as a detector as well. If the

* J.W. Strutt, Baron Rayleigh, § 269 of *The Theory of Sound*, 2nd ed. (Macmillan, London, 1894, reprinted by Dover Books, 1945).

† Actually in Lord Rayleigh's time the ultrasound was generated mechanically, using 'bird whistles', and doing experiments in air it was the velocity which was taken as known (to be the same as that of audio-frequency sound) so that by determining λ experimentally, the frequency could be found by using eqn. (1.5).

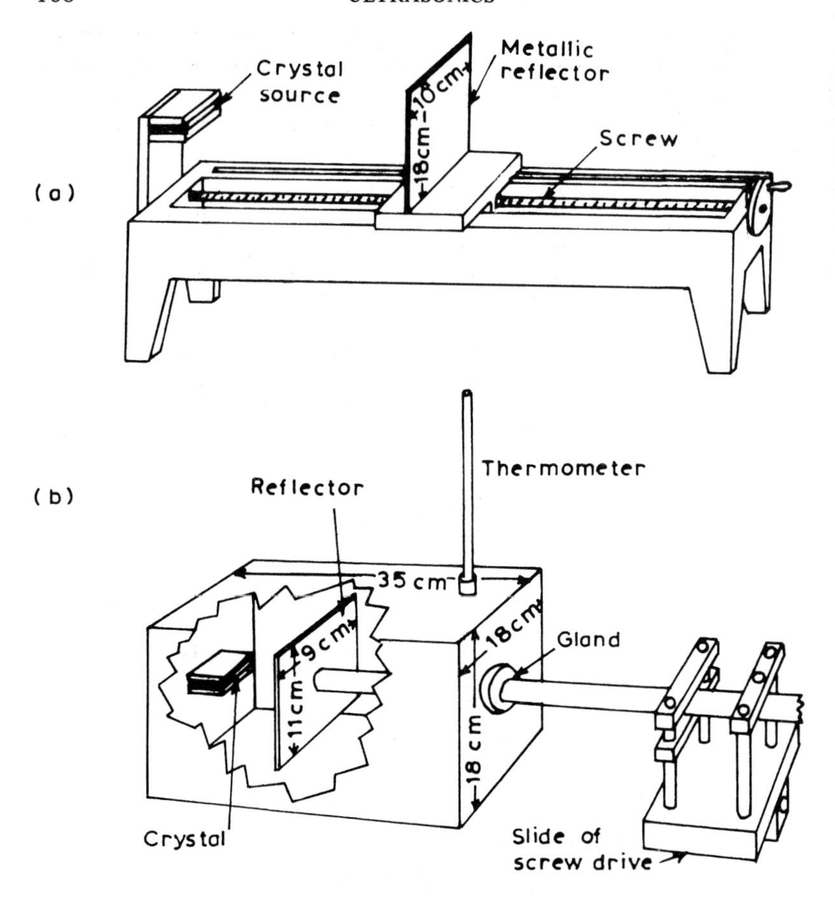

Fig. 7.2. Pierce's interferometer, (*a*) for use with air, and (*b*) for use with other gases. (From G.W. Pierce, 1925 *Proc. Amer. Acad. Sci.*, **60**, 271.)

anode current, indicated by the milliammeter in the circuit shown in fig. 4.9, is recorded as the reflector is moved, it will be found that this current is a periodic function of the position of the reflector (see fig. 7.3). Measurement of the distance between adjacent peaks in fig. 7.3 gives the value of $\frac{1}{2}\lambda$ and so, as with Kundt's tube measurements, if the frequency is known the velocity can be calculated. Variations on this particular standing-wave method can be used for ultrasonic measurements in liquids. Thus, instead of having a movable reflector one could propagate the ultrasound vertically downwards and use as the reflector the bottom of the vessel containing the liquid. Then the transducer will have to be mounted so that it can be moved up and down by measurable distances in a

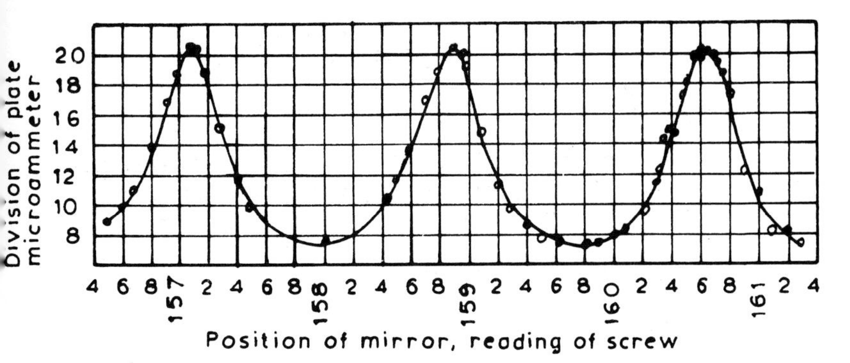

Fig. 7.3. Plot of readings of microammeter against readings of position of reflecting mirror (fig. 7.2). (From G.W. Pierce, 1925, *Proc. Amer. Acad. Sci.*, **60**, 271.)

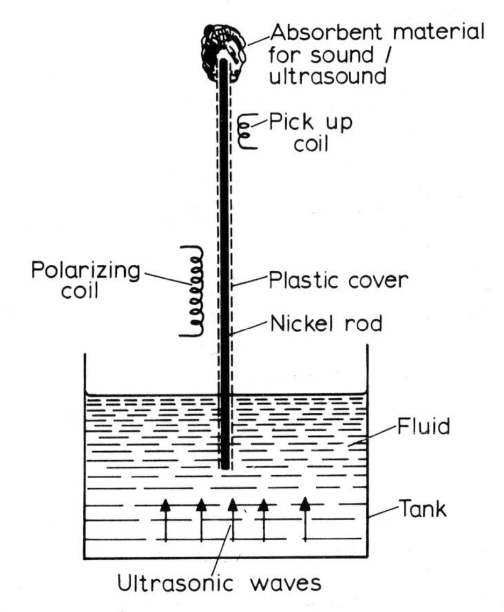

Fig. 7.4. Apparatus for generating stationary waves in a liquid. (From J. Blitz, *Fundamentals of Ultrasonics* (Butterworths, London 1967).)

controlled manner. In this circumstance it might also be convenient to replace the piezoelectric transducer by a magnetostrictive transducer in the form of a nickel rod, for example, see fig. 7.4.

In liquids there is another rather interesting way of detecting the standing ultrasonic wave pattern and measuring the ultrasonic wavelength. It was noted in Sections 1.2 and 2.3 that the excess pressure $p(x, t)$, varies along the path of an ultrasonic wave

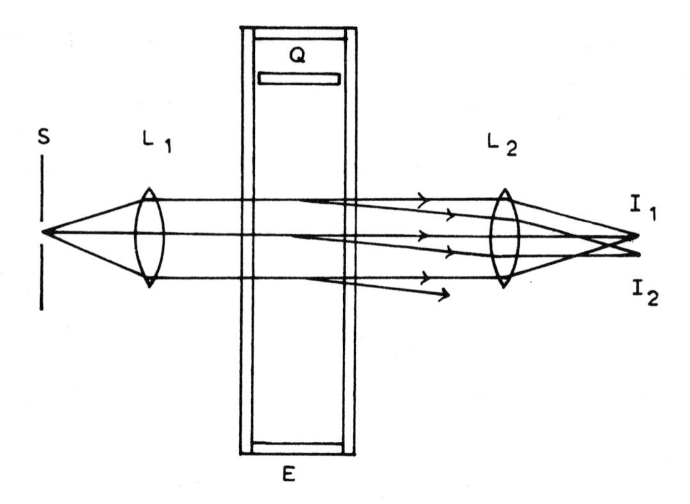

Fig. 7.5. Arrangement used by Debye and Sears for the observation of optical diffraction by ultrasonic waves.

$$p(x, t) = -ikEu(x, t). \tag{2.51}$$

Since the medium, in this case a liquid, is compressible there will be corresponding variations of the density of the liquid. These variations in the density, in turn, mean that if the liquid is transparent there will be corresponding variations of the refractive index along the path of the ultrasonic wave. The realization of this led to the idea that it might be possible to use the standing ultrasonic wave pattern in a liquid as an optical diffraction grating. The successful detection of an optical diffraction pattern from stationary ultrasonic waves in a liquid was performed in 1932, independently, by Debye and Sears and by Lucas and Biquard. A sketch of the experimental arrangement used by Debye and Sears is shown in fig. 7.5. It is found that the effective grating spacing is λ_s, the ultrasonic wavelength. The reason why it is λ_s and not $\frac{1}{2}\lambda_s$ as one might perhaps have expected is somewhat subtle – see also the discussion of Bragg diffraction imaging in Section 7.4. The conventional diffraction condition for a bright diffracted beam from a 'one-dimensional' grating with spacing a, namely

$$a \sin \theta = n\lambda_o \tag{7.1}$$

becomes

$$\lambda_s \sin \theta = n\lambda_o \tag{7.2}$$

where λ_o is the optical wavelength. Thus bright beams occur at angles θ given by

$$\sin \theta = \frac{n\lambda_o}{\lambda_s} \tag{7.3}$$

where n is an integer. Thus from measurements of the optical diffraction pattern one can determine the ultrasonic wavelength in the liquid.

The standing-wave methods we have described are capable of yielding quite accurate values for ultrasonic velocities. It is also possible, but rather more difficult, to use them to determine attenuation coefficients. For example, if one uses Pierce's interferometer to determine the velocity, by determining the wavelength, one is only concerned with the peak separations in fig. 7.3. As the path length is increased, the height of the peaks will be decreased because of the extra attenuation suffered by the ultrasound along the longer path. However, if one is to obtain quantitative values of the attenuation coefficients it is necessary to make quite major alterations to the experimental arrangement and also to analyse the peaks quite carefully. The optical diffraction method can also be adapted to enable attenuation coefficients, instead of wavelengths, to be measured. This is done by measuring the intensity of the zero-order diffraction line using different parts of the ultrasonic beam as the diffraction grating, that is for part of the beam near to the generator and then for part of the beam further away from the generator. Better results will be obtained by using progressive waves rather than stationary waves, that is by eliminating the reflector. The fact that the ultrasound, and therefore the 'diffraction grating' is then moving will not be very important since the velocity of the ultrasound is very small in comparison with the velocity of light in the medium. This method of determining attenuation coefficients is capable of producing results of quite reasonable accuracy under favourable conditions. It is, of course, limited to transparent media.

(*b*) *Pulse-echo methods.* In our first discussion of pulse-echo methods in Section 5.1 we regarded the ultrasonic velocity c in sea water as known so that distances could be determined by using

$$d = \tfrac{1}{2}ct. \qquad (5.1)$$

However, it is also possible to perform this type of pulse-echo experiment in a medium in which the ultrasonic velocity is unknown but in which d can be measured with a metre rule, calipers, a micrometer, etc. Then, from the measurement of t, using the oscilloscope trace of the echo pattern (see fig. 5.7) c can be calculated using eqn. (5.1). Thus, whereas in standing-wave methods the ultrasonic velocity is determined via eqn. (1.5) from measurements of λ, in pulse-echo methods the velocity is determined from measurements of length and time. For ultrasonic velocity measurements in solids pulse-echo methods are probably much more popular than standing-wave methods, whereas for liquids and gases pulse-echo methods are much less likely to be used than one of the standing-wave methods.

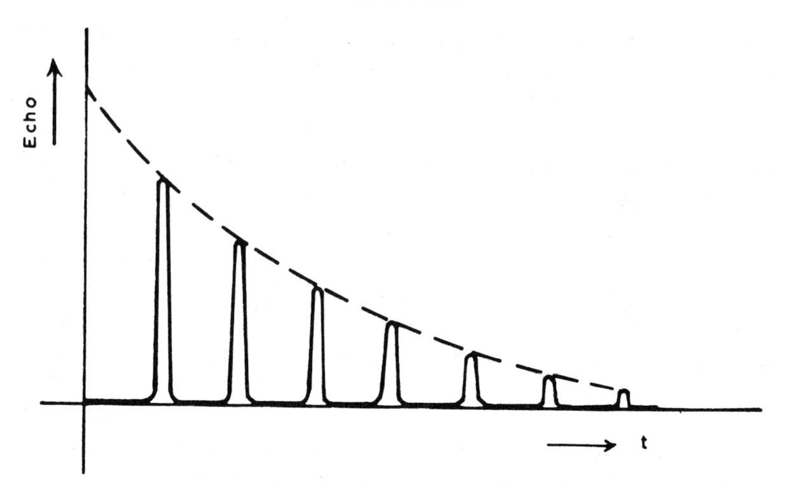

Fig. 7.6. A schematic diagram of multiple-echo pattern.

The determination of attenuation coefficients with pulse-echo techniques involves studying a multiple-echo pattern rather than using just a single echo. Thus, if we consider the situation illustrated in fig. 5.6(*a*), the first echo will be received after the ultrasonic pulse has made one complete round journey of length $2d$, where d is the thickness of the specimen. However, not all of the pulse will pass from the specimen back into the probe but will be reflected back into the specimen to make a second return journey through the specimen, and so on. Thus, instead of a single echo there will be a whole sequence of echoes received by the detector at times corresponding to integer multiples of $(2d/c)$. These echoes will be of decreasing amplitude (fig. 7.6). Therefore from measurements of the heights of successive peaks the value of α_p can be determined. In practice, some refinements are necessary if one wishes to obtain good values of attenuation coefficients.

7.2. *Image formation*

The ultrasonic pulse-echo techniques which we have already described could have been regarded as using ultrasound as a substitute for radiation in certain parts of the electromagnetic spectrum, in particular for radio waves and for X-rays. In a similar manner, this chapter can be regarded as being concerned with the possibility of the use of ultrasound as a substitute for radiation in the optical part of the electromagnetic spectrum. As in previous chapters we shall still be concerned with low-power applications.

There are both advantages and disadvantages in using ultrasound instead of light to 'see' or examine things. One important advantage lies in the fact that many media which are opaque to light are quite transparent to ultrasound. Also, of course, there is much greater freedom of choice of wavelength; whereas optical wavelengths are confined within rather narrow limits, one can obtain a much wider range of ultrasonic wavelengths simply by varying the ultrasonic frequency. Another possibility is that details in some specimens may be shown up better with ultrasound than with light. The disadvantages of ultrasound include the fact that not only is it invisible, but also it does not directly affect a photographic film or plate. One should be careful not to pursue the analogy between ultrasound and light too far. One reason is that sound, or ultrasound, requires a material medium in which to propagate, whereas electromagnetic waves can be propagated *in vacuo*. Another obstacle to the analogy lies in the fact which has just been mentioned, namely that whereas an optical image may be perceived by the human eye, an ultrasonic image cannot be perceived directly; it must, somehow, be converted into an optical image.

The study of optics can be divided into ray optics and wave optics. Ray optics includes the rectilinear propagation of light ('light travels in straight lines'), the reflection and refraction of light, and the exploitation of these properties in various optical instruments. Wave optics includes the phenomena of interference, diffraction and holography.

We have seen in Chapter 5 how the difficulty of the invisibility of ultrasound can be overcome by using 'time-of-flight' measurements of ultrasonic pulses. In the pulse-echo method there is no ultrasonic 'image' in a physical sense; any image is constructed electronically, e.g. by a B-scan. At the present time the pulse-echo method forms the basis of most of the successful systems for obtaining information from an ultrasonic investigation of a physical or biological system. We have mentioned some of these applications of the method in Chapters 5 and 6. However, this success of pulse-echo techniques may serve to obscure, rather than to emphasize, the analogy between ultrasound and light. For not only is the pulse-echo method analogous to ray optics rather than to wave optics, but it is also analogous to only a small part of ray optics, namely rectilinear propagation and simple reflection phenomena.

As we have seen already, the well established, man-made ultrasonic pulse-echo systems involve scanning an object, usually with piezoelectric transducers. Elaborate electronic gadgetry is then used to construct a visible picture from the scan. Instead of scanning with a transducer and using an electronic image-reconstruction system, it is also possible to render ultrasonic images visible in other ways that actually involve the use of visible light and are much closer, at least in the theory of their operation, to more conventional optical systems. Although the original work on some of these methods took place before the development of pulse-echo techniques, progress in developing technically and commercially viable systems was not at all rapid until very recently. In this chapter we shall discuss ultrasonic techniques that form alternatives to pulse-echo techniques. One possibility involves the production of an ultrasonic image, either as a shadow or by using an ultrasonic lens system. One then has to solve the problem of visualization of the ultrasonic image, that is converting it into an optical image; this may, or may not, be done holographically. Another possibility involves the production of an ultrasonic hologram by the interference of two coherent ultrasonic beams, one of which encounters an object while the other is a reference beam. The ultrasonic hologram is then converted into an optical hologram; this hologram can then be used for the production of an optical image. So far the technology is mostly at the research and development stages; it has not yet reached the stage at which these ultrasonic methods can compete successfully with the pulse-echo method and its associated electronic visualization systems, while there is no reason to suppose that they would ever replace an optical instrument or system in a situation in which the optical instrument could be used (i.e. with transparent media).

In figs 7.7 and 7.8 we illustrate schematically the important ways of forming an image. These diagrams are not, at this stage, to be regarded as being concerned with any particular physical kind of rays or waves. Thus, they could be relevant to the illumination of an object with light, or with X-rays or radio waves, or with ultrasound, or even with beams of particles such as electrons or neutrons. For light and for some other kinds of waves it is possible to construct lenses, but there are some kinds of waves for which it is much more difficult, or even impossible, to construct lenses. The system in fig. 7.7 involves no lenses and is, at first sight, simpler than that in fig. 7.8 which involves using a lens.

The examination of the bones or internal organs of a patient or of an animal using X-rays provides an example of the use of the system in fig. 7.7 in its transmission mode. Here the X-ray wavelength is very much smaller than the dimensions of the object that is being examined and so the resolution of detail is not a serious problem in this case (the question of resolution will be discussed in more detail in Section 7.3). For light, the wavelength is of the order of a thousand or so times larger than a typical X-ray wavelength and optical images formed by the system in fig. 7.7 are likely to be too blurred to be useful. However, for light the system of fig. 7.7, which does not involve lenses, can be adapted to give images without too much blurring if holographic techniques are used. This involves using a second beam which bypasses the object and interferes with the beam which is transmitted by the object to produce a hologram which can then be recorded on a photographic plate. An image can subsequently be

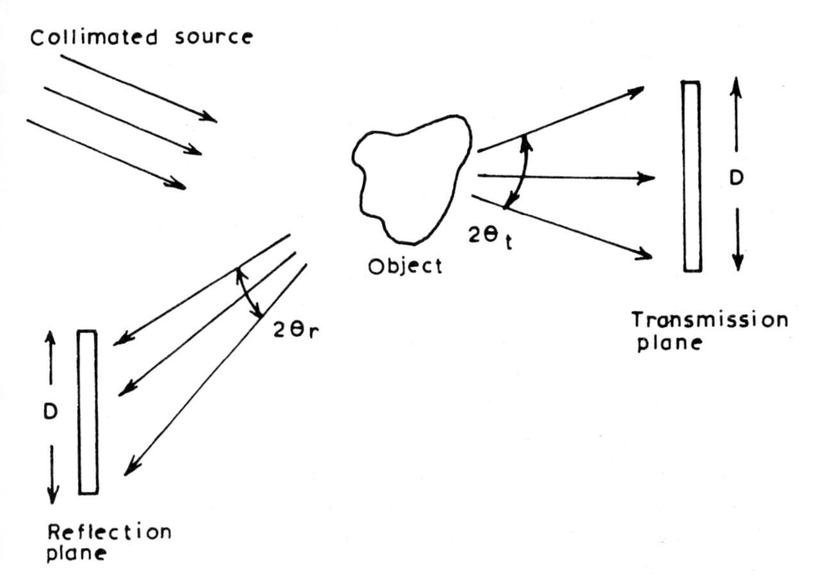

Fig. 7.7. Image formation without a lens.

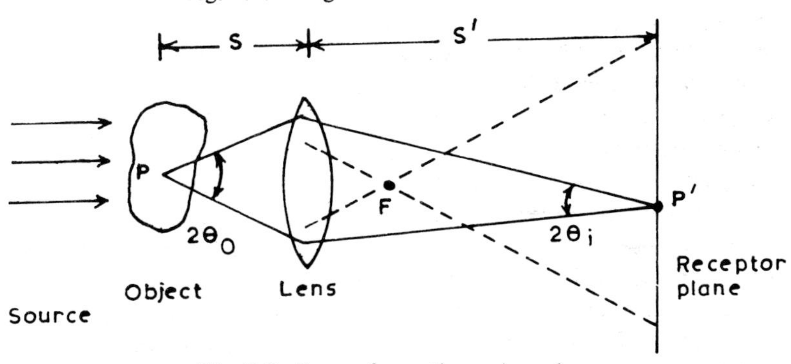

Fig. 7.8. Image formation using a lens.

reconstructed from the hologram. Although holography has, so far, been exploited most successfully at optical wavelengths, the principles of holography apply to any kind of waves. Thus, one could use two coherent ultrasonic beams to produce an ultrasonic hologram and this has indeed been done. The ultrasonic hologram can be converted into an optical hologram, i.e. on a photographic plate, and then the image can be reconstructed by using a laser in the conventional manner. A further discussion of ultrasonic holography will be found in Section 7.5.

There are often, however, advantages in using lenses if this is possible. Therefore the system in fig. 7.8 is probably more familiar to us than that in fig. 7.7 because we are more familiar with the formation of images with light rather than with other kinds of waves. Thus, very many optical instruments can be found which rely for their operation on image-formation by a lens or by a system of lenses. Consequently, some ultrasonic systems have been designed involving the use of ultrasonic lenses. However, the design and construction of an ultrasonic lens is not quite as trivial a task as one might, at first sight, expect. Good resolution depends, loosely speaking, on making the ratio of the wavelength of the radiation to the diameter of the lens as small as possible (see Section 7.3). For light and using a glass lens of diameter 60 mm this ratio is about $(6000 \times 10^{-10})/(6 \times 10^{-2}) = 10^{-5}$. For an ultrasonic wave of, say, velocity $1000\,\mathrm{m\,s^{-1}}$ and frequency 1 MHz, $\lambda = c/\nu = 10^3/10^6\,\mathrm{m} = 10^{-3}\,\mathrm{m}$ and so for a lens of diameter 60 mm the ratio is $10^{-3}/6 \times 10^{-2}) \doteqdot 1{\cdot}7$. This is about five orders of magnitude larger than for the optical lens. Although one can make an ultrasonic lens slightly larger, say of diameter 0·5 m, the resolution will clearly still be very poor compared with the resolution achieved with optical lenses. The ratio could be improved by reducing the ultrasonic wavelength (i.e. by increasing the frequency) but the higher the frequency used, so also the higher will be the attenuation of the ultrasound. Thus one is only likely to use ultrasonic lenses in situations in which it is not possible to use optical lenses, for example in cameras for use in deep murky water where there is little light available, such as when searching for the monster in Loch Ness.

As we have already hinted, the very success of ultrasonic pulse-echo techniques led to a considerable neglect of work on the development of ultrasonic lenses. The problems that have to be overcome in the design of an ultrasonic lens include poor acoustic matching between the lens and the medium in which the ultrasound is propagating. Thus with a solid lens immersed in water the match is very poor and this leads to substantial losses. Also, mode conversion at the surface of the lens can lead to shear-wave propagation in the lens and to a second focus. If one uses a liquid lens, the acoustic matching can be improved substantially and also the absorption will generally be lower than with a solid lens. However the liquid lens needs to be contained in some shell. If the shell is made of a soft plastic or thin rubber to reduce the losses, the shape of lens will be determined by the elastic properties of the material of the shell and by the pressures within and without the lens. To impose a fixed shape on a liquid lens needs a rigid, and therefore thicker, shell and this will lead to greater losses than with a thin and flexible shell. Another difficulty is that a good lens will need a large aperture; this involves either making the lens thicker, which leads to more absorption, or having a combination of lenses, which leads to more losses due to acoustic mismatch at the additional lens

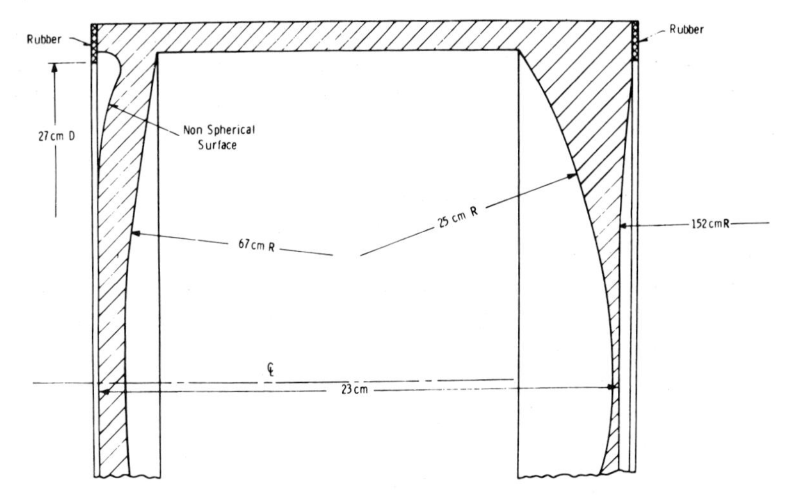

Fig. 7.9. Cross-section of a doublet polystyrene lens used in a sonic camera recently developed by Westinghouse for real time imaging in turbid water. (From C.H. Jones and G.A. Gilmour, 1976, *J. Acoust. Soc. Am.* **59**, 74.)

surfaces. Recently there has been some revival of interest in the development of lenses for use in sonic cameras for underwater work (see fig. 7.9).

7.3. *Resolution*

In our discussion of the pulse-echo method in the previous two chapters we avoided any detailed or quantitative discussion of the question of resolution. Although we have chosen to discuss resolution in this chapter on ultrasonic 'optics', it is important not to seek too close an analogy between the cases of ultrasonic waves and visible light in this context. For optical instruments such as telescopes and microscopes it is diffraction which is the most important factor in limiting the resolution of the instrument. Thus the angular limit of resolution α for a telescope objective is given by

$$\alpha = \frac{1{\cdot}22\lambda}{a} \tag{7.4}$$

where λ is the wavelength and a is the aperture of the objective of the telescope. A similar formula can be obtained for the limit of resolution h of a microscope objective, that is the smallest distance between two objects that can be resolved,

$$h = \frac{1{\cdot}22\lambda}{2n \sin U} \tag{7.5}$$

where $2U$ is the angle subtended at one of the objects by a diameter of the objective lens, and n is the refractive index. (The acoustic microscope will be discussed in Section 7.4) Both these formulae are based on the Rayleigh criterion applied to the diffraction by the objective lens. Equations (7.4) and (7.5) are both examples of the more general formulae involving the limitation of resolution by diffraction, namely

$$\text{Limit of resolution} = \frac{1{\cdot}22\lambda}{2(\text{NA})} \tag{7.6}$$

where (NA) denotes the *numerical aperture*. This type of formula can also be applied to the limit of resolution for an ultrasonic imaging system if it is diffraction that is the most important factor limiting the resolution.

In optical microscopes and telescopes, diffraction is very important in limiting the resolution that can be achieved; several other factors, such as the bandwidth, that is the spread of frequencies present in the light, and the finite length of the optical wavetrain are relatively much less important. In the case of optical instruments the basic problems in resolution are generally that an object may be too small or too far away to be seen, or that the angular separation between two objects may be too small for their images to be able to be distinguished from one another. On the other hand, in ultrasonic methods in general, and in the pulse-echo method in particular, these are not always the greatest problems. The amount of reflected ultrasonic energy that arrives at the detector may be too small to be distinguished from the 'noise' background, or there may be difficulty in obtaining a sufficiently accurate location of an object from the analysis of the reflected ultrasound because of the restricted duration of the ultrasonic pulse. These factors may prove to be more serious in practice than the limit to the resolution imposed by diffraction.

In ultrasonic pulse-echo methods one has deliberately chosen to use a pulse, that is a relatively short wavetrain, and so the pulse duration may affect the resolution that can be achieved. Suppose that a pure sinusoidal ultrasonic (or any other) wavetrain with a single wavelength λ, or frequency ν, is 'chopped' into pulses. If one of these pulses is Fourier-analysed it will be found to contain not only the frequency ν but a spread of frequencies around ν (see fig. 7.10). The shorter the pulse the greater will be the spread of frequencies. The 'bandwidth' is a measure of this 'spread' of frequencies. For a 'square' pulse the bandwidth is $1/\tau$ where τ is the duration of the pulse.

Suppose that we are concerned with using pulse-echo techniques not just to detect an object but also to identify its position, or possibly its velocity, as accurately as possible. In the pulse-echo method it is, essentially, the transit time t from transmission of the pulse to detection of the echo that is measured and so the range d can be found from

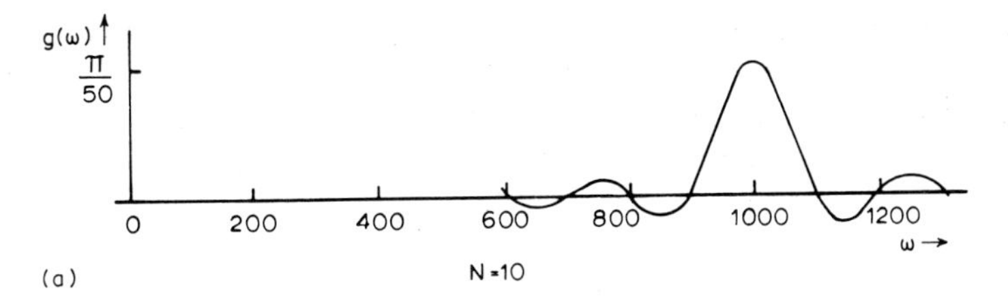

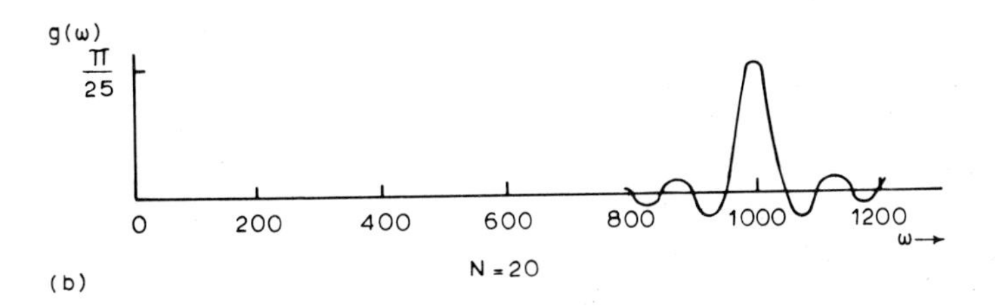

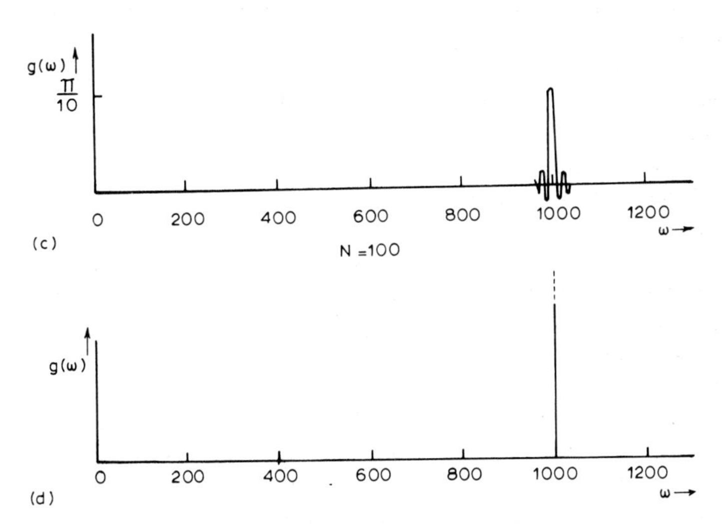

Fig. 7.10. Fourier transform $g(\omega)$ of finite sinusoidal 'gated' wavetrains with $\omega_0 = 1000\,\text{rad}\,\text{s}^{-1}$ with pulse lengths of N wavelengths for (*a*) $N = 10$, (*b*) $N = 20$, (*c*) $N = 100$, and (*d*) an infinite wavetrain of the same frequency.

$$d = \tfrac{1}{2}ct. \tag{5.1}$$

There will be an uncertainty, or error, in the measured time interval t and therefore in the range calculated from t. If E is the total signal energy received in the echo and N_0 is the noise power per unit bandwidth, then it is possible to show that

$$\delta t = \frac{1}{\beta}\sqrt{\left(\frac{N_0}{2E}\right)} \tag{7.7}$$

where δt is the smallest change in t that can be detected and β is the bandwidth. Using eqn. (5.1) we see that the corresponding uncertainty, δd, in the range is then given by

$$\delta d = \tfrac{1}{2}c\delta t = \frac{c}{2\beta}\sqrt{\left(\frac{N_0}{2E}\right)}. \tag{7.8}$$

If we take $\beta = 1/\tau$ (see above) for a square pulse then

$$\delta d = \frac{c\tau}{2}\sqrt{\left(\frac{N_0}{2E}\right)}. \tag{7.9}$$

This means that the use of a shorter pulse length will give a better range-resolution. There will be a limit to the accuracy that can be achieved by reducing the pulse length more and more, because the pulse energy will also be reduced – unless the amplitude is correspondingly increased. An alternative way to increase β in order to improve the range resolution is to sweep the frequency of the ultrasound through a range $\Delta\nu$ during the pulse. Provided $\tau\Delta\nu$ is large the bandwidth β of the frequency-swept pulse will be approximately equal to $\Delta\nu$ so that

$$\delta d = \frac{c}{2\Delta\nu}\sqrt{\left(\frac{N_0}{2E}\right)}. \tag{7.10}$$

We have already mentioned in Section 6.2 the use of frequency-swept pulses by many species of bats in echo-location.

In addition to δd, the uncertainty in the range which we have just considered, one can also consider $\delta\theta$, the uncertainty in the direction, which is given by

$$\delta\theta = \frac{\lambda}{l}\sqrt{\left(\frac{N_0}{2E}\right)} \tag{7.11}$$

where l is the effective aperture. Alternatively, if one is concerned with using the Doppler effect to determine the velocity of a target, rather than its instantaneous position, there will be an uncertainty, or error $\delta(\Delta\nu)$, in the frequency shift given by

$$\delta(\Delta\nu) = \frac{1}{\tau}\sqrt{\left(\frac{N_0}{2E}\right)}. \tag{7.12}$$

Therefore, using eqn. (2.36) for reflection from a moving target, we see that δv, the error in the velocity of the target, is given by

$$\delta v = \frac{c}{2\nu\tau} \sqrt{\left(\frac{N_0}{2E}\right)}. \tag{7.13}$$

Thus the accuracy of a velocity determination will be improved by increasing the value of $\nu\tau$, which is the number of cycles in each pulse. A comparison of eqns. (7.9) and (7.13) shows that to obtain accurate range resolution requires a small value of τ, the pulse duration, but accurate velocity determination requires a large value of τ. Equation (7.13) implies that for a given frequency ν the best velocity determination will be achieved by having a large value of τ, the pulse duration, i.e. a narrow bandwidth. Equation (7.9), however, shows that to obtain accurate range resolution will require a small value of τ, i.e. a large bandwidth. Thus a desire to achieve maximum accuracy in range determination and a desire to achieve maximum accuracy in velocity determination impose opposite requirements on the bandwidth. The choice of τ, the duration of each ultrasonic pulse generated in an ultrasonic pulse-echo system, will therefore be governed by the purpose for which it is intended to use the system.

It may be that one is concerned with using ultrasonic techniques for detection, without making highly accurate determinations of the position or velocity of an object. Then the important consideration is the power of the ultrasonic signal that reaches the detector. The power that is received at a detector for a given output of power from the ultrasonic generator will depend on, among other things, the geometrical configuration used in the system. The two extreme cases are those of a 'point source', when the transmitter radiates power uniformly in all directions, and of a 'plane-wave' beam of ultrasound generated by the transmitter. In practice neither of these ideal situations will occur exactly, but very good approximations to these geometries may be obtained in many cases. Thus, for example, in the detection of submarines one can regard the transmitter as a point source so that at a distance r the transmitted power P_t is spread out over the surface of a sphere of area $4\pi r^2$ (see fig. 7.11). The power crossing each unit area of surface of this sphere will be $P_t/4\pi r^2$; if σ is the effective cross-section of the target the amount of power scattered by the target will be $\sigma P_t/4\pi r^2$. If it is assumed that this power is scattered by the target uniformly in all directions too, the power returning to unit area of the detector placed near the source will be

$$P_r = \left(\frac{\sigma P_t}{4\pi r^2}\right) \Big/ 4\pi r^2 = \frac{\sigma P_t}{16\pi^2 r^4}. \tag{7.14}$$

This equation, which is sometimes known as the *radar equation* because it was applied to radar before it was applied to sonar, shows that, for a given receiver sensitivity in range-finding by sonar, the transmitted power will

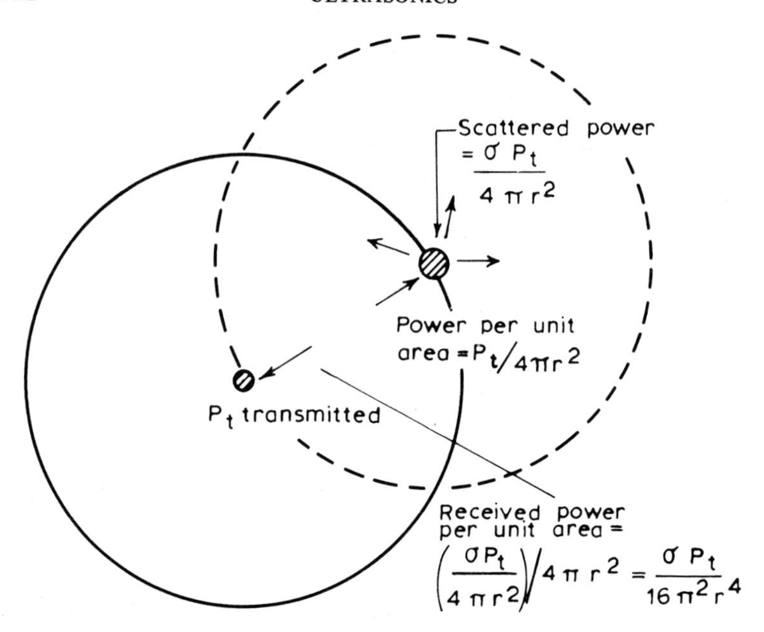

Fig. 7.11 Diagram to illustrate the 'radar equation'.

need to vary as the fourth power of the required range. Thus to double the range will require 16 times the transmitted power, that is an increase of +12 dB.

The effective cross-section σ of a target that appears in eqn. (7.14) is not necessarily the same as the geometrical cross-section. Figure 7.12 illustrates the variation of σ for a spherical target of radius a as the relative magnitudes of the wavelength and of the size of the target are varied. From this figure we see that σ is the same as the area of cross-section if $\pi d \gg \lambda$. There is an oscillatory region for $\lambda \lesssim \pi d \lesssim 10\lambda$ and finally for very small spheres σ is very much smaller than the area of cross-section of the spheres; this is the Rayleigh scattering region.

In non-destructive testing, whether in medical or industrial applications, we are much closer to the 'parallel beam' situation. The problems of obtaining a signal that is large enough to detect, from a crack in a casting or from some tissue variations in a human body, will depend to some extent on whether we are using the reflection or transmission mode of operation (see fig. 7.7). If we have a detector situated in the field of the reflected ultrasound, the power reflected by the discontinuity that we are trying to detect in the medium will depend on the acoustic mismatch at the discontinuity (see eqn. (2.44)). Generally speaking, the larger the acoustic mismatch the larger will be the reflected power for a given power

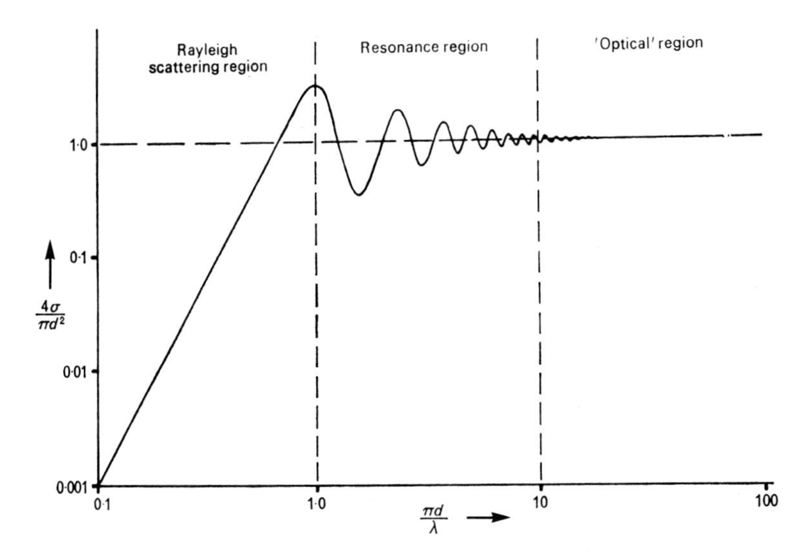

Fig. 7.12 The relationship between the echo-reflecting power of a sphere and the wavelength of the transmitted signal. (From G. Sales and D. Pye, *Ultrasonic Communication by Animals* (Chapman and Hall, London, 1974).)

in the incident beam. But the acoustic mismatch at a discontinuity within a specimen is something that it is often not within our control to alter. If a detector is placed in the field of the transmitted ultrasound, it is likely that the amount of ultrasonic energy scattered by the crack, defect, or discontinuity under investigation, will be only a small fraction of the incident energy. We then have the problem of distinguishing this scattered ultrasound from the remainder of the ultrasound which passes through the specimen unscattered. It is possible to overcome this difficulty by using a lens with a 'zero-order stop' placed at its focal point. The unscattered ultrasound will be brought to a focus at the focal point and stopped while the ultrasound scattered at P will be focused to form an image at P' (see fig. 7.13).

7.4. *Visualization*

Whether one uses a system with lenses or a system without lenses, there is always the problem that at some stage an ultrasonic image has to be 'visualized', or converted into an optical image. Historically, the first attempts to convert an ultrasonic image into an optical image were made using light; this was before the days of the electronic developments that made the pulse-echo technique feasible. For many years the possibility of

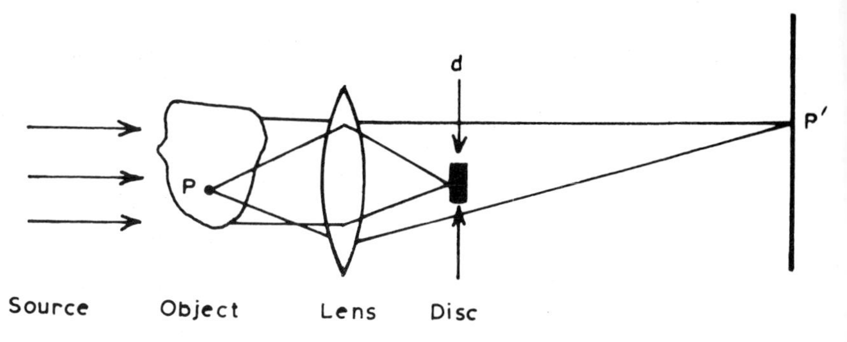

Fig. 7.13 Use of small opaque disc as a 'zero-order stop'.

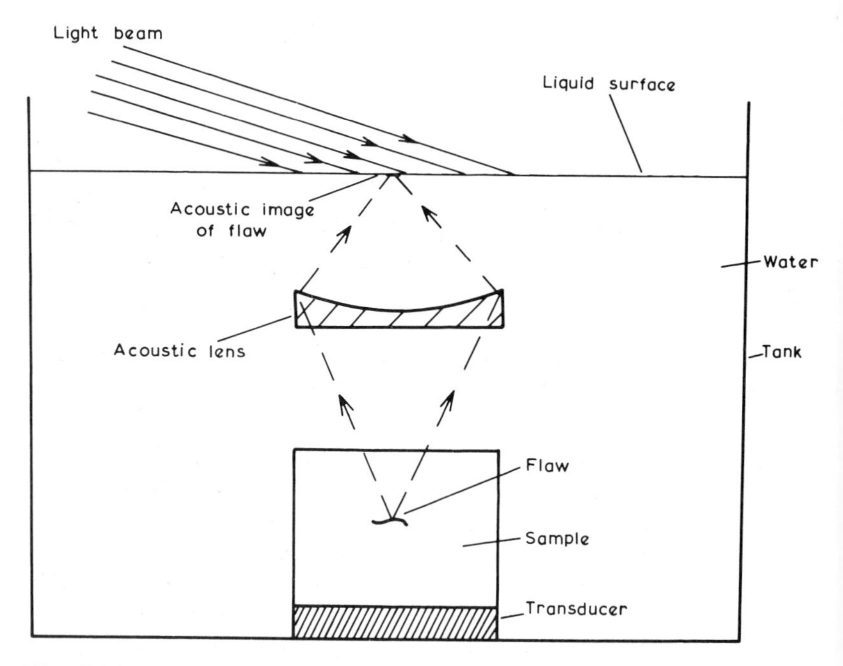

Fig. 7.14. An early flaw-detection system devised by Sokolov in 1934. (From J. Blitz, 1972, *Phys. Bull.*, **23**, 23.)

optical visualization was very much over-shadowed by the successful development of the electronic visualization schemes used with the pulse-echo technique (see Chapter 5). More recently other methods have been developed which are more closely related to optics.

Figure 7.14 illustrates an early flaw-detection system devised by Sokolov in 1934. The acoustic image is formed as a ripple pattern at the surface of the liquid and the image is viewed with light incident obliquely

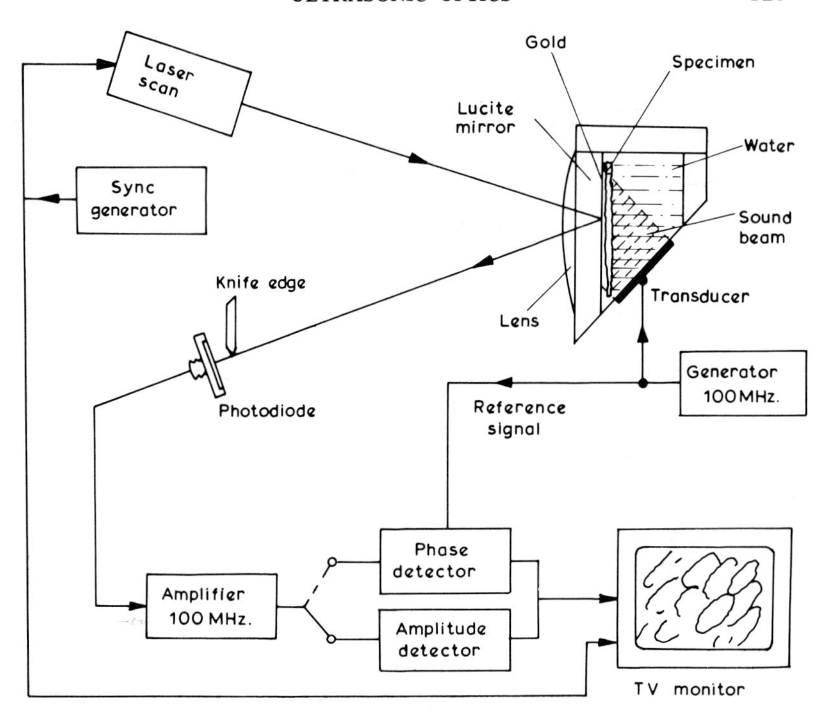

Fig. 7.15 Principle of acoustic (ultrasonic) microscope. (From A. Korpel, L.W. Kessler, and P.R. Palermo, 1971, *Nature*, **232**, 110.)

on the surface. Nowadays a laser would be used to scan the ripple pattern and the reflected light would be fed back into an appropriate system of electronics to reproduce an image on a television screen. This scheme has been used in the construction of a system known as an *ultrasonic microscope* or sometimes an *acoustic microscope* (fig. 7.15). In recent years this has progressed from being merely an idea and has passed through the research and development stage to become a commercial proposition. Figure 7.16 shows an example of results obtained from an acoustic microscope compared with those obtained from optical and electron microscopes. An ultrasonic microscope may, in due course, come to be regarded as complementary to an optical microscope, with each being able to distinguish certain features of a specimen which would remain indistinguishable with the other, and with the ultrasonic microscope perhaps being less destructive to the specimen than the electron microscope.

The mention of the use of a laser scan of an ultrasonic image or ripple pattern to produce a visualization of that image electronically suggests the possibility of using the laser scan to produce a hologram of that ultrasonic

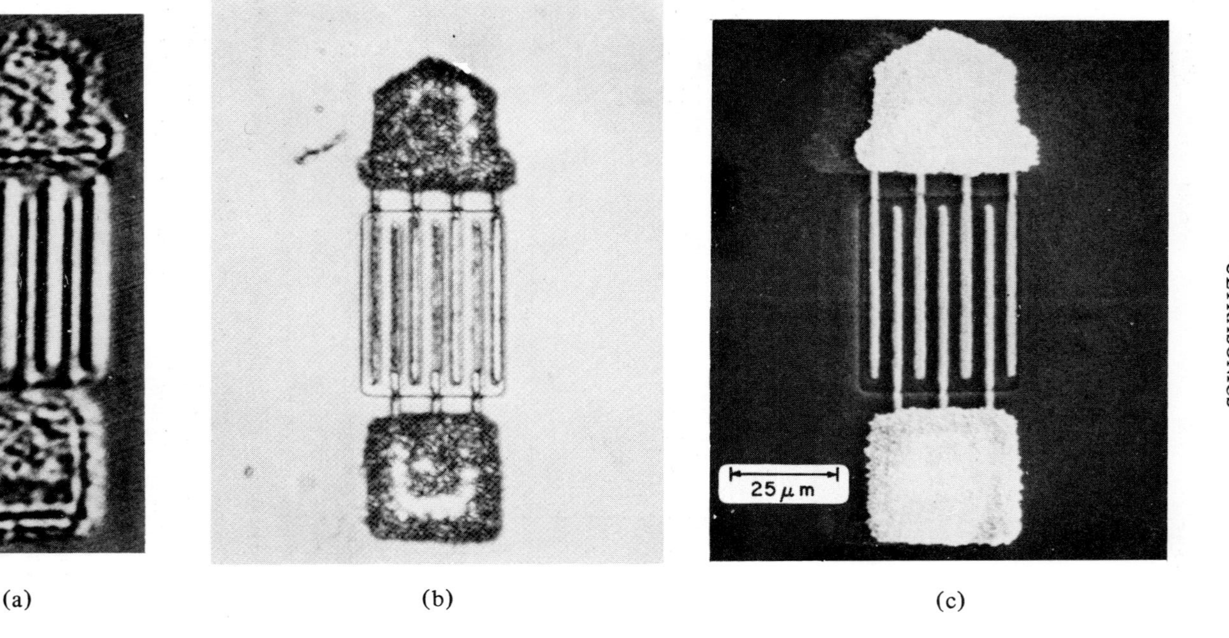

Fig. 7.16. Images of a bipolar transistor obtained with (*a*) acoustic microscope, (*b*) optical microscope and (*c*) electron microscope (From R.A. Lemons and C.F. Quate, 1974, *Appl. Phys. Letters*, **25**, 251.)

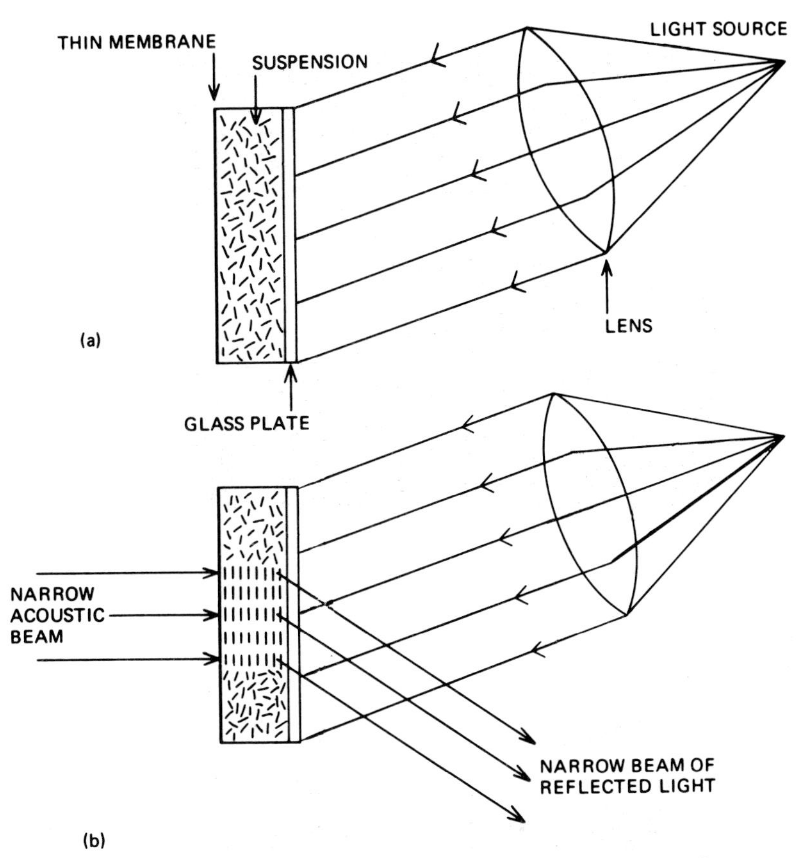

Fig. 7.17. Diagrams illustrating Pohlmann cell, (*a*) no ultrasound, (*b*) in use.

image. This can then be followed by an optical reconstruction of the ultrasonic vibration pattern from the hologram, which constitutes a direct application of *optical holography*, in which the ultrasonic ripple pattern is regarded as the object. It is important not to confuse this with what is normally meant by the term *ultrasonic holography* or *acoustic holography*.

There are alternatives to forming an ultrasonic image as a ripple pattern. For example, one can direct the ultrasound onto a Pohlmann cell (fig. 7.17). One face of the cell is acoustically transparent and the opposite face, of glass, is optically transparent. The cell contains a suspension of fine aluminium flakes in some suitable liquid. Ordinarily, the metallic flakes will be randomly oriented but when ultrasound traverses the suspension the metallic flakes tend to align themselves perpendicular to the direction of propagation of the ultrasound. The degree of orientation,

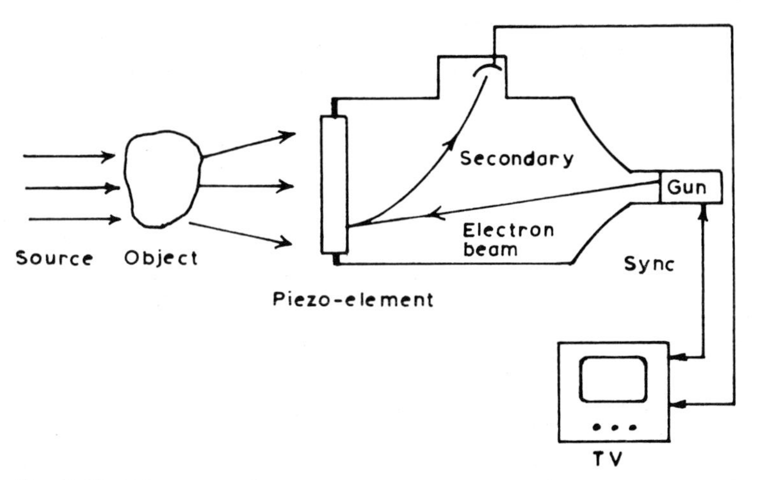

Fig. 7.18. Diagrammatic representation of system using Sokolov tube.

which gives an indication of ultrasonic intensity, can be detected by observing light reflected by the suspension (see fig. 7.17(*b*)).

Another method for image conversion is to use a large quartz transducer as a screen on which the ultrasonic image may be received. Quartz is an insulator and therefore, if the crystal has no metallic electrode attached to its surface, the charges which are produced will remain on the parts of the screen where they are generated and the charge distribution will reproduce the ultrasonic vibration pattern. The image, in the form of a charge distribution on the back of the quartz screen can be visualized by scanning the screen with an electron beam in much the same way that the charge distribution produced by the optical image in a TV camera is scanned. The image can then be reconstructed on a television screen. The system is sketched in fig. 7.18.

Another possible method for visualization is to use Bragg diffraction. We have already mentioned in Section 7.1 the method of Debye and Sears for determining the ultrasonic velocity by establishing a standing ultrasonic wave and using the standing wave as an optical diffraction grating. In a somewhat similar manner it is possible to obtain optical diffraction patterns from the refractive-index variations that arise in a transmitted ultrasonic beam. The reconstruction of an optical image of the original object from such a diffraction pattern constitutes the basis of Bragg-diffraction systems of ultrasonic visualization; a block diagram of such a system is shown in fig. 7.19(*a*). An important difference between the Debye–Sears system and the Bragg-diffraction systems is that in the Debye–Sears system the diffraction is caused by a stationary ultrasonic wave pattern,

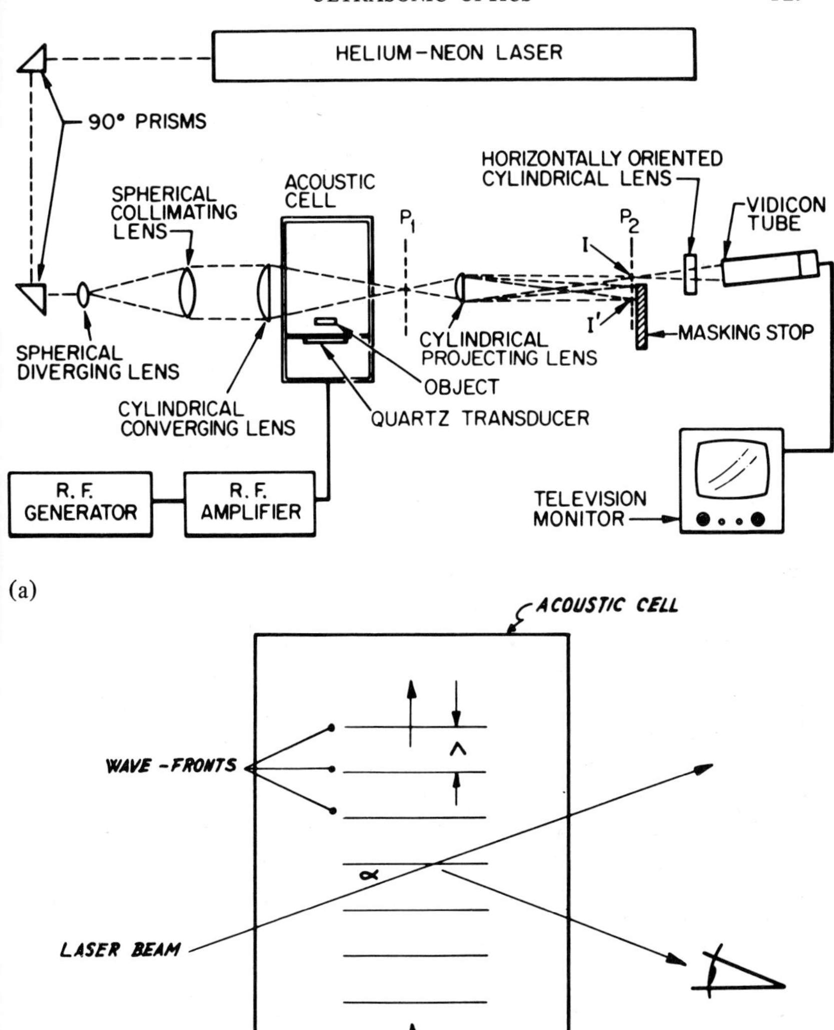

Fig. 7.19. (*a*) Diagram of a Bragg-diffraction imaging system, (*b*) acoustic cell.

whereas in the Bragg-diffraction system the diffraction is caused by a propagating ultrasonic wave.

To try to understand Bragg-diffraction imaging consider the diffraction of a plane monochromatic light wave, of wavelength λ, by a (progressive) plane monochromatic sound wave or ultrasonic wave, of wavelength Λ, The acoustic cell, without an object in it, is represented in fig. 7.19(*b*). We suppose that the width of the ultrasonic beam is very great in comparison with λ. The ultrasonic wavefronts are analogous to the planes of atoms in a crystal in which Bragg-diffraction of X-rays occurs. For an arbitrary angle α between the light beam and the ultrasonic wavefronts there will be destructive interference between light waves diffracted at different positions across the ultrasonic beam and so no diffracted light beam will be seen at this angle. However, for certain special angles the diffracted waves will all reinforce one another to give a bright diffracted light beam. The diffraction condition for a bright beam is analogous to the Bragg-diffraction condition for X-rays

$$\sin \alpha = \frac{n\lambda}{2d_{\mathrm{hkl}}} \tag{7.15}$$

but the crystal-plane spacing d_{hkl} is replaced by the distance between successive ultrasonic wavefronts, that is by the ultrasonic wavelength Λ, so that

$$\sin \alpha = \frac{n\lambda}{2\Lambda}. \tag{7.16}$$

n is an integer and it is reasonable to suppose that, after the direct beam, the strongest diffracted beam will be for $n = 1$. Once an object is placed in the acoustic cell, the transmitted ultrasonic beam is no longer a simple plane wave but a complicated wave pattern; this complicated pattern could be regarded as a Fourier synthesis of a large number of plane waves with different wavelengths and consequently with different Bragg-diffraction angles.

7.5. *Ultrasonic holography*

Holography was developed, or invented, by D. Gabor in the late 1940s. The most striking feature of holography is that it enables one to store a three-dimensional image, in the form of a hologram, on a two-dimensional photographic plate. The image may be reconstructed by viewing the developed plate, as a transparency, in coherent light. Although holography was first developed using visible light, Gabor showed that the principle of holography is applicable to any wave motion. Therefore, it should be possible to form a hologram using sound, or ultrasonic waves, instead of visible light.

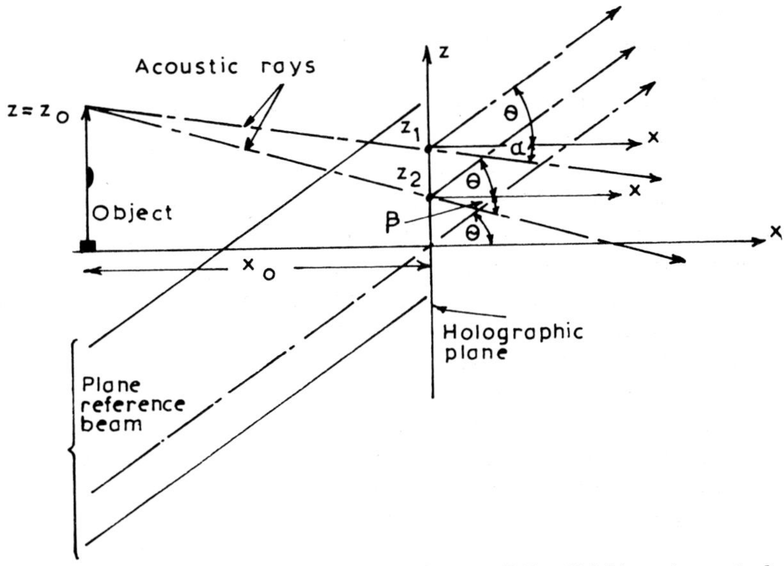

Fig. 7.20. Formation of a hologram. (From E.E. Aldridge *Acoustical Holography*, Merrow, Watford, 1971.)

A hologram is formed by the interference of a reference beam and a beam transmitted by an object (see fig. 7.20). At the plane $x = 0$ the displacement associated with the reference beam, which is assumed to be plane, can be written as

$$u_1(x, y, z, t) = a_1 \exp\ [\mathrm{i}(\omega t - k \sin \theta z) \tag{7.17}$$

where $k = 2\pi/\lambda$; we write this as

$$u_1(x, y, z, t) = a_1 \exp\ (\mathrm{i}\psi_1) \tag{7.18}$$

for short. The beam from the top of the object is not, strictly, a parallel beam, but it can be approximated as such; therefore at $x = 0$ the displacement associated with this beam can be written as

$$u_2(x, y, z, t) = a_2 \exp\ \{\mathrm{i}[\omega t + \phi(z_0)] + k \sin \alpha z\} \tag{7.19}$$

which we can write as

$$u_2(x, y, z, t) = a_2 \exp\ (\mathrm{i}\psi_2) \tag{7.20}$$

for convenience. The combined effect produced at the plane $x = 0$ when the beams interfere is then given by

$$a \exp\ (i\psi) = a_1 \exp\ (\mathrm{i}\psi_1) + a_2 \exp\ (\mathrm{i}\psi_2). \tag{7.21}$$

The sum of these two complex numbers is found as shown in fig. 7.21.

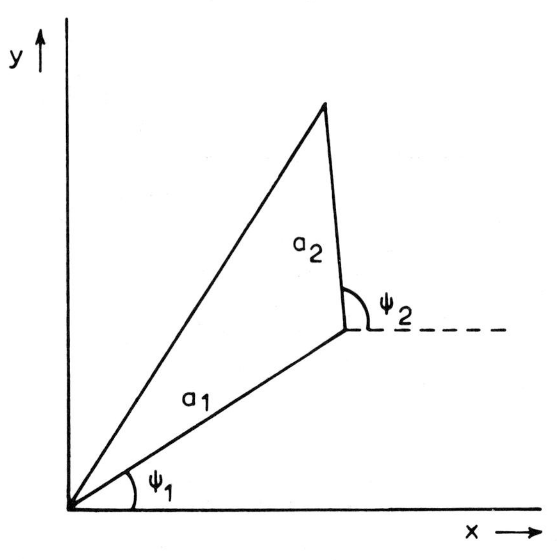

Fig. 7.21. Diagram to illustrate the addition of two complex numbers $a_1 \exp(i\psi_1)$ and $a_2 \exp(i\psi_2)$.

The amplitude, a, of the displacement at $x = 0$ is then given by

$$\begin{aligned} a^2 &= a_1^2 + a_2^2 - 2a_1a_2 \cos(\psi_1 + \pi - \psi_2) \\ &= a_1^2 + a_2^2 + 2a_1a_2 \cos(\psi_2 - \psi_1). \end{aligned} \tag{7.22}$$

Let us consider the intensity which is proportional to a^2 and let us suppose that we have some 'black box' device that will convert ultrasonic intensities directly into optical intensities. The importance of considering optical intensities, rather than amplitudes, lies in the fact that the blackening of a photographic plate is, within certain limits, proportional to the intensity of the light, i.e. is proportional to a^2. a^2 can then be regarded as containing two terms, the first $(a_1^2 + a_2^2)$ is independent of z and constitutes a uniform background, while the second term $2a_1a_2 \cos(\psi_2 - \psi_1)$ does vary with z, since

$$2a_1a_2 \cos(\psi_2 - \psi_1) = 2a_1a_2 \cos[\phi(z_0) + k(\sin\theta + \sin\alpha)z]. \tag{7.23}$$

It is the photographic record of this intensity distribution which constitutes a hologram. (We have, of course, simplified the picture by considering only rays from one point on the object and by suppressing the y coordinate in the plane $x = 0$.)

To reconstruct an image, the hologram is illuminated with coherent light (with a wavelength which is likely to be very much smaller than the

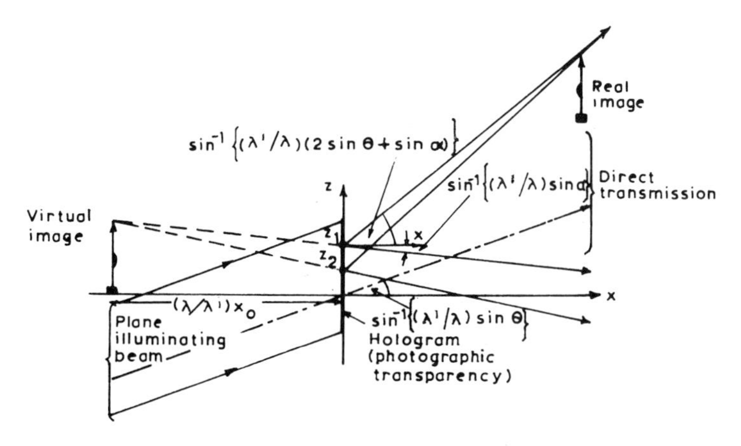

Fig. 7.22. Formation of an image from a hologram. (From E.E. Aldridge *Acoustical Holography*, Merrow, Watford, 1971.)

ultrasonic wavelength) (see fig. 7.22). We suppose that this too is a plane-wave beam which may be represented by

$$u_r(x,y,z,t) = \exp[i(\omega' t - k' \sin\theta' z)] \tag{7.24}$$

where $\omega'/2\pi$ and k' are the frequency and wave vector of the light, and θ' is the angle between its direction of propagation and the x axis. The angle θ' should be chosen so that $k \sin\theta = k' \sin\theta'$ or $(2\pi/\lambda)\sin\theta = (2\pi/\lambda')\sin\theta'$ so that

$$\theta' = \sin^{-1}\left(\frac{\lambda'}{\lambda}\sin\theta\right). \tag{7.25}$$

After passing through the hologram the amplitude of this beam will be modulated by the intensity variations recorded in the hologram, thus the transmitted amplitude will be proportional to

$$\cos[\phi(z_0) + k(\sin\theta + \sin\alpha)z] \times \exp[i(\omega' t - k' \sin\theta' z)]. \tag{7.26}$$

Recalling that we imposed the condition $k' \sin\theta' = k\sin\theta$ and taking only the real part of this expression we have

$$\begin{aligned} &\cos(\omega' t - k\sin\theta z)\cos[\phi(z_0) + k(\sin\theta + \sin\alpha)z] \\ &\propto \cos[\phi(z_0) + k(\sin\theta + \sin\alpha)z + \omega' t - k\sin\theta z] \\ &+ \cos[-\phi(z_0) - k(\sin\theta + \sin\alpha)z + \omega' t - k\sin\theta z] \\ &= \cos(\omega' t + \phi(z_0) + k\sin\alpha z) + \cos(\omega' t - \phi(z_0) - k(2\sin\theta + \sin\alpha)z). \end{aligned} \tag{7.27}$$

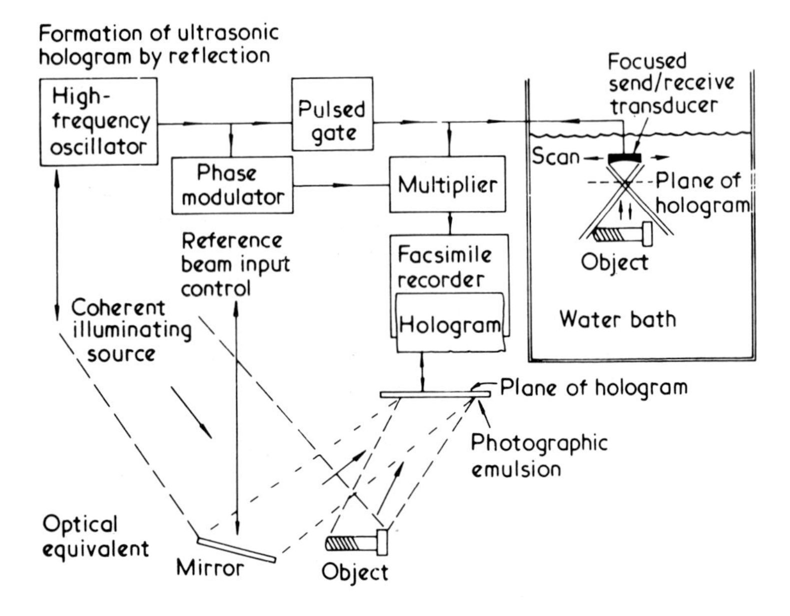

Fig. 7.23. Formation of ultrasonic holograms. (From E.E. Aldridge, A.B. Clare, and D.A. Shepherd, 1974, *Ultrasonics*, **12**, 155.)

The first term on the right-hand side of eqn. (7.27) is of the same form as the real part of the right-hand side of eqn. (7.19). Therefore this term corresponds to a virtual image of the original real object (see fig. 7.22). It is possible to show that the second term corresponds to a real image; if the angle between this beam and the x axis is θ'' then $k' \sin \theta'' = k(2 \sin \theta + \sin \alpha)$, that is

$$\theta'' = \sin^{-1} \left(\frac{\lambda'}{\lambda} (2 \sin \theta + \sin \alpha) \right). \tag{7.28}$$

In the above account we have ignored the details of the 'black box' process that is needed for converting the acoustic hologram into an optical hologram. This conversion is far from being a trivial operation; a system for achieving it is outlined in figs. 7.23 and 7.24. The ultrasonic hologram may be scanned with a focussed transducer (see fig. 7.23) and the output from this detector is then processed electronically to produce an optical hologram, either on the screen of an oscilloscope or as a permanent record on a photographic film or plate (see fig. 7.24). In the example shown in these two diagrams the object is a flaw in a block of metal.

Acoustical holography is usually performed at kHz or MHz frequencies, with the aim of investigating objects immersed in a fluid such as water or for investigating ordinary sized objects that are opaque to visible light.

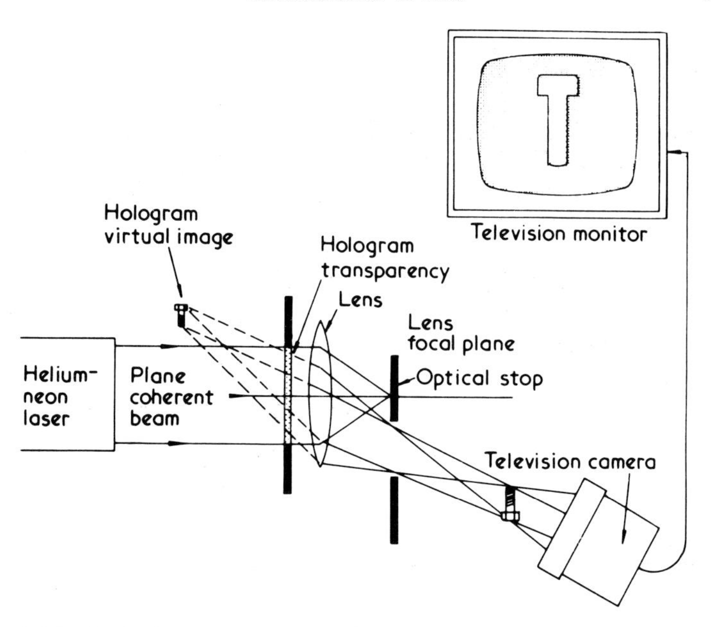

Fig. 7.24. Optical reconstruction of ultrasonic holograms. (From E.E. Aldridge, A.B. Clare, and D.A. Shepherd, 1974, *Ultrasonics,* **12**, 155.)

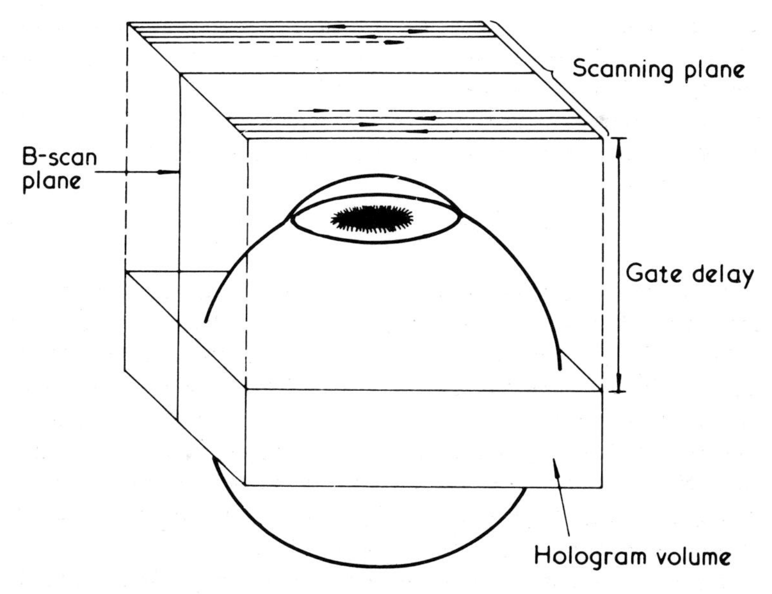

Fig. 7.25. B-scan and hologram geometry, (From R.C. Chivers, 1974, *Ultrasonics,* **12**, 209.)

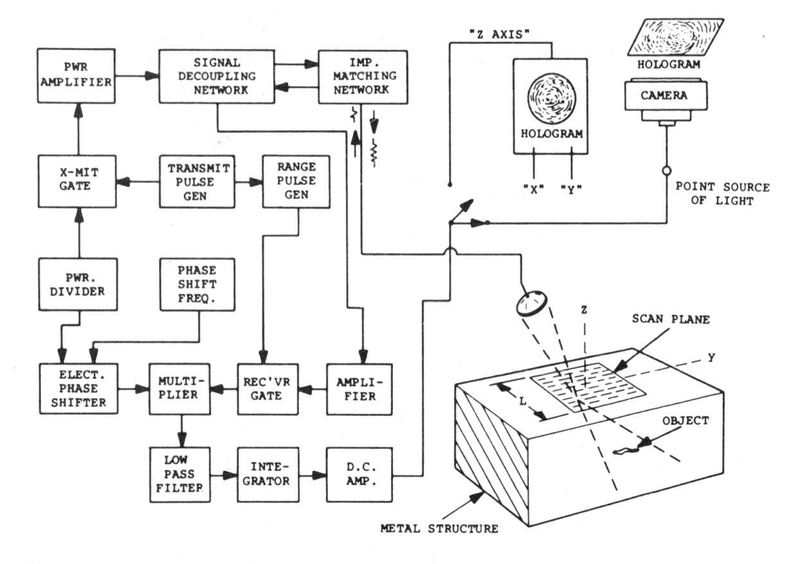

Fig. 7.26. Electronic block diagram of a scanned acoustical holography system. (From G. Wade, Ed., *Acoustic Imaging* (Plenum, New York, 1976).)

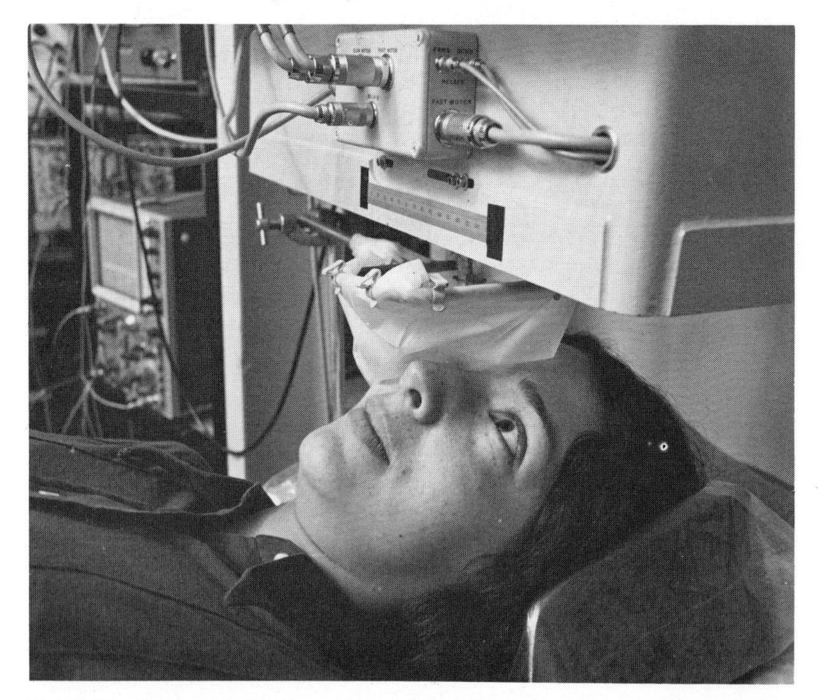

Fig. 7.27. Arrangement of patient and fluid bath for ultrasonic holographic examination of the eye (Photograph by courtesy of E.E. Aldridge and UKAEA.)

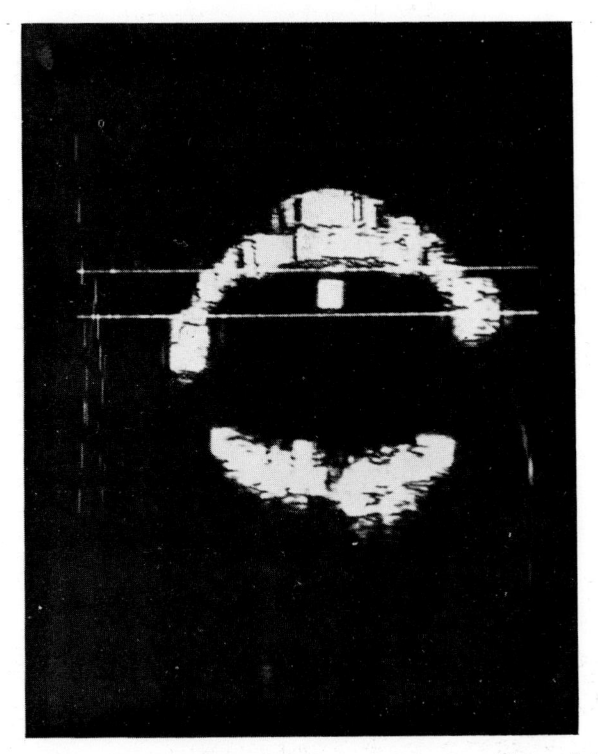

Fig. 7.28. Linear B-scan of excised eye showing position of hologram acceptance gate. (From R.C. Chivers, 1974, *Ultrasonics,* **12**, 209.)

Ultrasonic holography could be used instead of optical holography in situations in which the object would be damaged by a laser beam but would be unharmed by ultrasonic waves. However, it now seems likely that the principal immediate uses of ultrasonic holography may be as an alternative to the pulse-echo method in such areas as medical and ophthalmic diagnosis, non-destructive testing, and nuclear reactor surveillance. In ophthalmology, for example, ultrasonic holography promises to become complementary to the more well-established pulse-echo B-scanning technique. To keep the signal-to-noise ratio at an acceptable level a hologram is constructed of only a relatively thin section of the eye. This section is perpendicular to the plane that is studied in a B-scan; the geometry of this is illustrated in fig. 7.25 for a supine patient looking straight up (see figs. 7.26 and 7.27). Some results are shown in figs. 7.28 and 7.29.

Another interesting possibility, since acoustical methods have been used for many years in seismological work, is to investigate whether holographic acoustical methods can be exploited in seismology. In order to achieve the

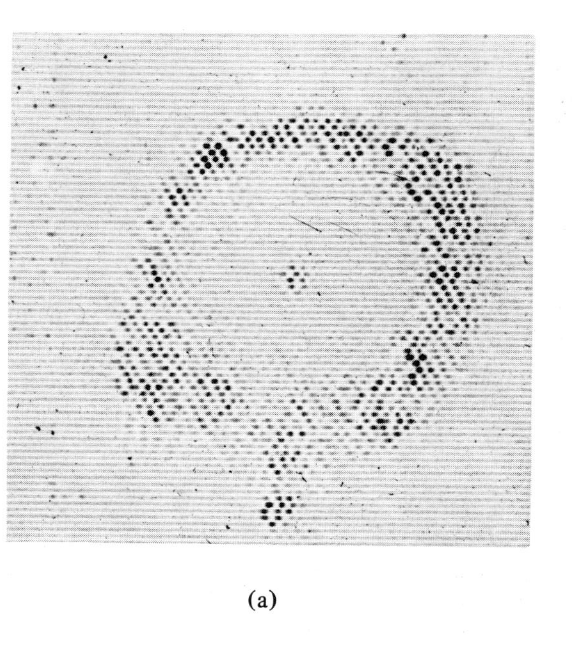

(a)

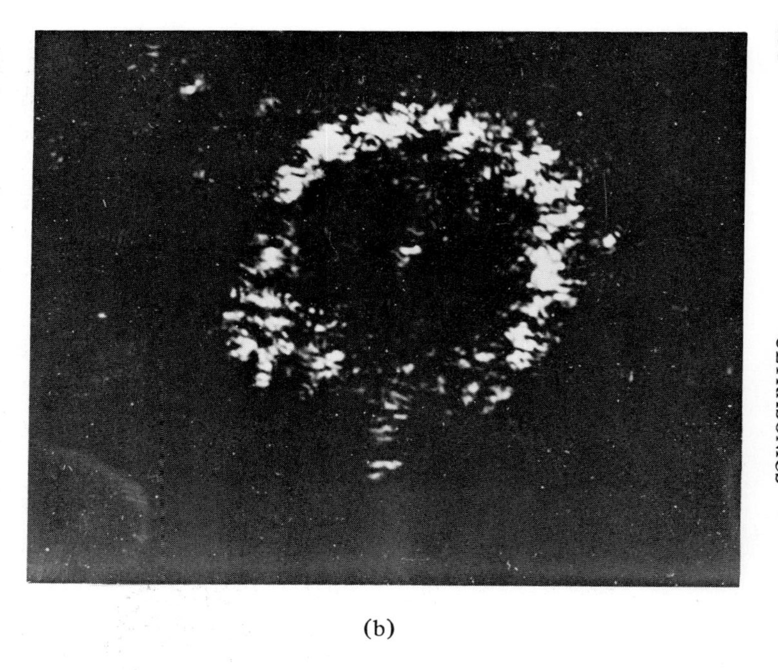

(b)

Fig. 7.29. (*a*) Hologram and (*b*) reconstruction of excised eye of fig. 7.28. The central bright spot is the back of the lens. (From R.C. Chivers, 1974, *Ultrasonics*, **12**, 209.)

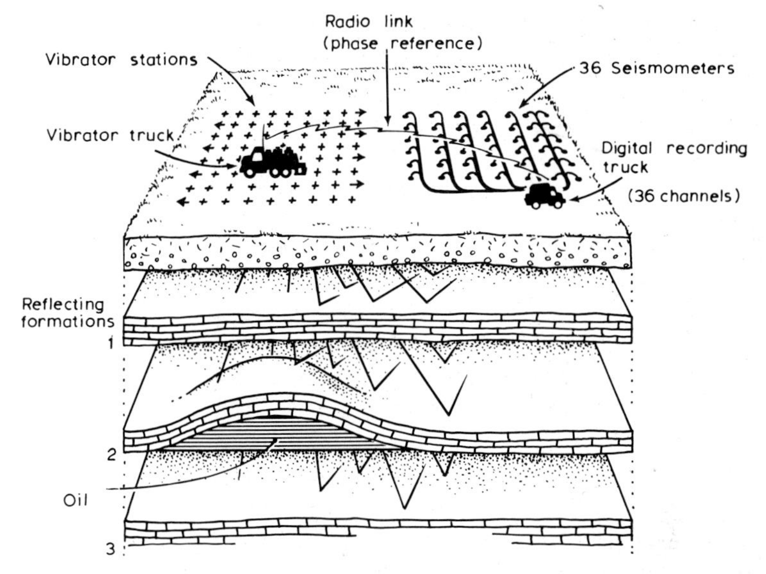

Fig. 7.30. Proposed earth holography method. (From A.F. Metherell and L. Laramore, Eds., *Acoustical Holography* Vol. 2 (Plenum, New York, 1970).)

necessary path lengths in the Earth for this work it is necessary to use very low frequencies, say 10–100 Hz, that is frequencies in the audio, or even sub-audio, range. Although such frequencies are, strictly speaking, outside the scope of this book it is interesting to note that the possibility of making Earth holograms has been investigated seriously (see fig. 7.30). The fact that the wavelengths at very low audible frequencies are so very much longer than at typical ultrasonic frequencies means that the wavelength ratio λ'/λ would be too small for optical reconstruction to be feasible. Instead, therefore a digital-computer method of reconstruction has been investigated.

7.6. *Acoustic emission*

Acoustic emission differs in a rather fundamental way from most other examples of the use of ultrasound. In this phenomenon there is no external source of ultrasound; instead, the specimen is subjected to stress and sound, or ultrasound, may be emitted at some point within the specimen, for example as the result of the growth of a crack. The sound or ultrasound that is generated can be detected by a transducer attached to the specimen and the signal from the transducer can then be fed into a suitable amplification and visualization system. We should hardly be surprised that the sound or ultrasound that is generated in acoustic emission is not just of one frequency but there is a wide range of frequencies present. The relationship between the nature of the pulse of sound or ultrasound that is generated and the deformation of the material is far from simple. There are different ways of processing the signal, such as 'ringdown counting' (threshold crossing counting), 'amplitude distribution analysis' and 'frequency analysis'. Each acoustic emission event in the specimen produces a pulse in the detecting transducer and ringdown counting simply involves counting these pulses. Amplitude distribution analysis involves measuring the amplitude of each signal generated by the transducer in response to an acoustic emission event and then constructing a histogram of the distribution of pulses as a function of energy. Frequency analysis involves making a Fourier analysis of the pulses to determine the relative intensities of components of various frequencies. Since the sound, or ultrasound, has to pass through the specimen and to be converted into electrical energy in the transducer, there is no reason to suppose that the amplitude or frequency distributions of the electrical pulses will be directly proportional to the corresponding distributions in the acoustic, or ultrasonic, pulses generated in the specimen.

The analysis and interpretation of the signals produced in acoustic emission is not easy. Careful experimentation has shown that the emission is highly specific to a given material and depends not only on the chemical composition of a material but also on its thermal and mechanical histories. The purpose of studying the signals produced in acoustic emission it to obtain information either about the intrinsic nature of the deformation process in the material under examination or about the existence and nature of flaws in the material. Some work on acoustic emission was performed in the 1940s and 1950s, but it was in the 1960s that the potential importance of the phenomenon really came to be appreciated and a great deal of research work on this subject was performed. The practical importance is principally in connection with the detection of faults in engineering structures. The uses, or potential uses, of acoustic emission include providing a useful technique for detecting flaws and anticipating failure in a wide variety of engineering structures subjected to stress. Particularly important examples are in the inspection of nuclear reactor pressure vessels and of parts of spacecraft, but many other applications are also possible in less spectacular situations.

8. Ultrasound as a form of energy in industry and medicine

The previous three chapters have been concerned with the wave properties of ultrasound and the uses of these in transmitting information. In this chapter we shall be more concerned with ultrasound as mechanical vibrations of a material medium and with the energy of those vibrations. Ultrasound may be effective in a mechanical fashion by the agitation that it causes, or it may be effective as a source of energy that can be applied locally without too much unwanted heating in the vicinity. We are generally concerned here with higher power levels than in the applications considered in previous chapters.

8.1. *Cavitation*

Suppose that a bubble of radius r exists in a liquid through which ultrasonic waves are propagating. The bubble may consist either of the vapour of the liquid itself or of some gas which has been dissolved in the liquid. This bubble will be subject to the changes in pressure associated with the ultrasound and so the bubble will contract and expand as the excess pressure rises and falls (see fig. 8.1). At the extreme of the rarefaction half of the cycle the bubble has its maximum radius r_{max} and at the extreme of the compression half of the cycle it has its minimum radius r_{min}. It is possible to show that if the excess pressure amplitude of the ultrasound is sufficiently high, in other words if the intensity of the ultrasound is sufficiently high, and the initial radius of the bubble is less than a certain critical value r_0, the bubble will suddenly collapse during the compressional half of the cycle with the sudden release of a comparatively large amount of energy. This collapse and the associated release of energy which occurs is known as *cavitation*. The critical radius r_0 is given by

$$\omega^2 r_0^2 = \left(\frac{3\gamma}{\rho}\right)\left(p_0 + \frac{2T_s}{r_0}\right) \tag{8.1}$$

where

$\gamma = c_p/c_v$

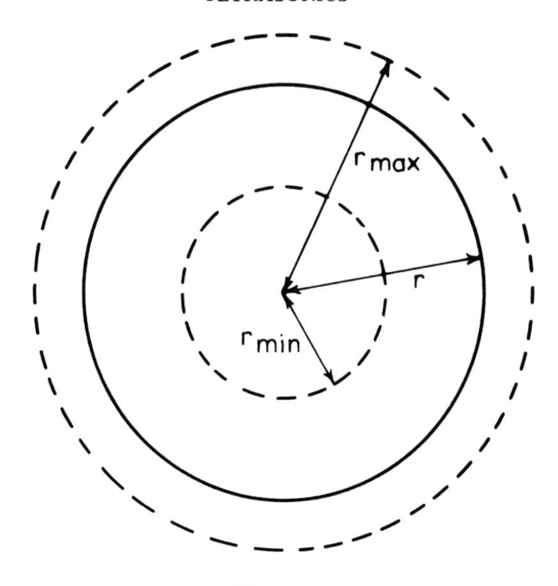

Fig. 8.1.

p_0 = the hydrostatic pressure (i.e. the pressure in the absence of ultrasound)
$\omega/2\pi = \nu$ = frequency of ultrasound
ρ = density of liquid
T_s = surface tension of liquid.

The pressure in a bubble just before it finally collapses may be very large indeed. Thus, when the bubble finally collapses an extremely powerful shock wave is produced and it is the energy in this shock wave which is responsible for many of the effects arising from cavitation. For example, a piece of metal placed in a liquid in which cavitation is occurring may become seriously pitted or eroded; this is usually referred to as *cavitation erosion*. The use of the term 'cavitation' has changed; in the early days it was used for what is now described as 'cavitation erosion', while the term 'cavitation' is nowadays used just to refer to the collapse of the bubbles and the associated release of energy.

The amount of energy released in cavitation depends on the ratio of r_{max}/r_0. If the intensity of the ultrasound is increased, this will mean that the amplitude of the oscillations of the excess pressure will be increased and this, in turn, means that the ratio r_{max}/r_0 will be increased, and so will the energy released in cavitation as an individual bubble collapses. If one considers cavitation in various different liquids, the energy released will be greater for larger values of the surface tension at the

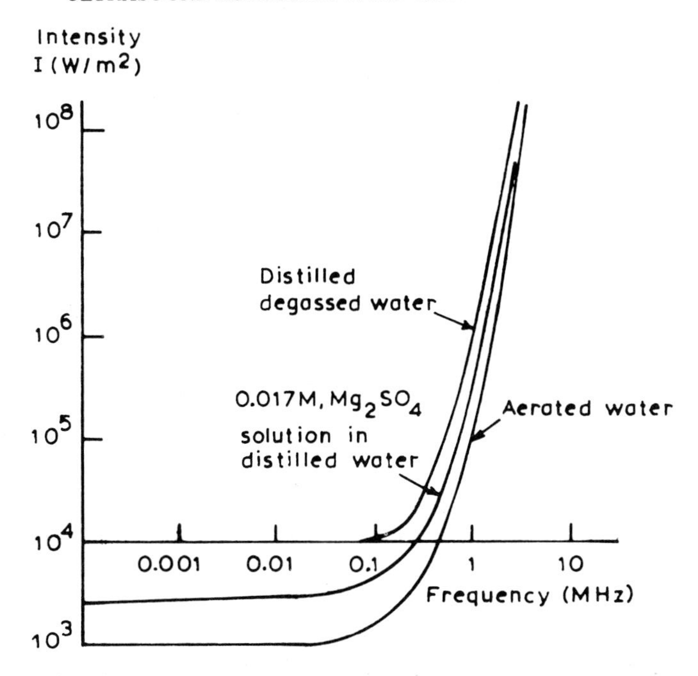

Fig. 8.2. Dependence on frequency of intensity threshold for cavitation in various aqueous media. (From R. Esche, 1952, *Acustica,* 2, AB, 208.)

surface of the bubble and for smaller values of the vapour pressure. Thus water, which has a high surface tension will involve large cavitational energies. The cavitational energies can be increased still further by adding some alcohol, say 10 per cent. This increases the vapour pressure, which increases the cavitational energies, although this is partly, but not completely, offset by the accompanying reduction of the surface tension.

Although the presence of bubbles produced by the release of a dissolved gas facilitates the onset of cavitation, it is also possible for cavitation to occur in gas-free liquids if the excess pressure amplitude of the ultrasound exceeds the hydrostatic pressure in the liquid. In part of the rarefaction half of the cycle the total pressure, $p_0 + p$, would become negative and so, to prevent this from happening, bubbles of vapour of the liquid will form. These bubbles will then collapse in the compressional half of the cycle giving rise to cavitation. For a given liquid and for a given ultrasonic frequency, there will be a certain minimum ultrasonic intensity needed to produce cavitation. This threshold intensity for the onset of cavitation varies with frequency (see fig. 8.2). Cavitation plays an important role in several of the applications of ultrasound that we shall discuss in the rest of this chapter.

8.2. *Ultrasonic cleaning*

Some work on the development of the idea of using ultrasonic cleaning was performed in Germany during the Second World War. Since that time ultrasonic cleaning equipment has been developed by many different manufacturers for a wide variety of applications.

Ultrasonic cleaning is achieved by immersing an article in a suitable cleaning fluid and passing ultrasound into the fluid. The cleaning is likely to be more efficient if both agitation and cavitation are involved. The shock waves produced during cavitation will reach any solid surface that is in the liquid and will scour or scrub that surface, with less probability of scratching or otherwise damaging a specimen than would occur in more conventional scrubbing. Moreover, by using ultrasound it is possible to take the scrubbing action into all the 'nooks and crannies' that would be inaccessible with a mechanical scrubbing brush. Also there are certain industries, such as the aerospace industry, in which normal cleaning methods cannot give the degree of cleanliness that is required for some of the products. In addition to the various specialist uses, the application of ultrasonic cleaning in quite ordinary industrial contexts may achieve a saving of time, labour, and cost.

In most ultrasonic cleaning applications, in order to ensure that cavitation occurs, one uses a frequency for which the threshold intensity for the onset of cavitation is quite low. From fig. 8.2 we see that this means using a frequency that, by ultrasonic standards, is quite low. For the comfort of the operator, however, it is desirable to use frequencies above the audible range (see also Chapter 10 on safety). Therefore, in practice, most ultrasonic cleaning equipment operates at frequencies in the region of 20 kHz. However, in the cleaning of very delicate articles, which might be damaged by cavitation, one uses higher frequencies (100 kHz to 1 MHz) so that the threshold intensity for the onset of cavitation is not exceeded. For most ultrasonic cleaning work standard commercial tanks, which are usually made of stainless steel, are quite suitable; the capacity ranges from about 200 ml to 100 litres (figs. 8.3 and 8.4), and they may be fitted with a system to circulate the cleaning fluid through a filter. If the cleaning fluid is toxic or inflammable (or both) it will be necessary to prevent the vapour from escaping from the tank.

The cleaning fluid varies from one application to another. It should not be regarded merely as an inert medium for the support of the ultrasonic vibrations and (if appropriate) cavitation; it should have the properties that one would expect of any good cleaning fluid used in a conventional cleaning process. Tap water, various acidic, alkaline and aqueous solutions, as well as some organic liquids, are used. In some cases the use of ultrasound

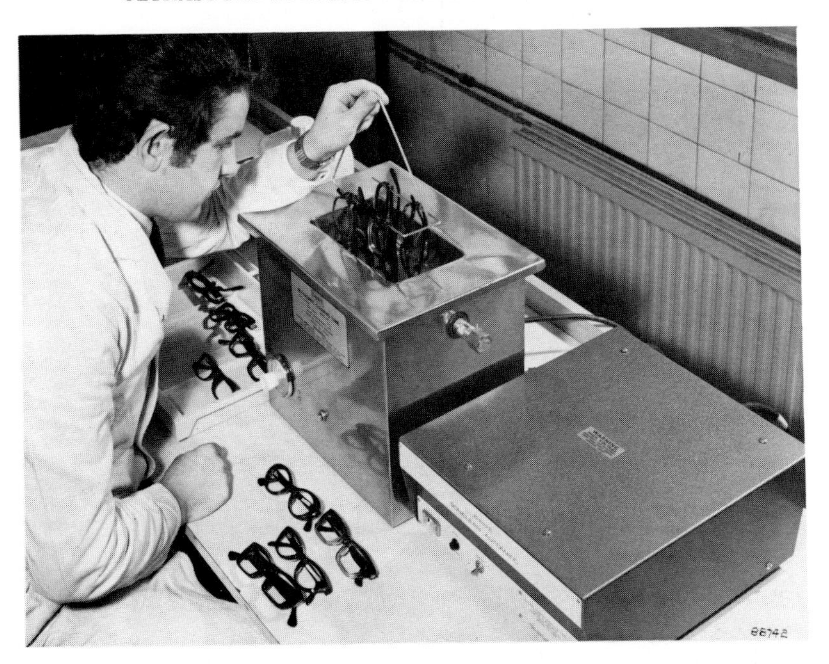

Fig. 8.3. Small ultrasonic generator and cleaning tank (Photograph by courtesy of Dawe Instruments Ltd.)

has enabled a relatively expensive solvent to be diluted or to be replaced by some other cheaper one.

The different types of source used in ultrasonic cleaning include mechanical generators, magnetostrictive transducers and piezoelectric transducers, although mechanical generators are no longer widely used. The electronic generator that drives the transducer is normally fairly simple, usually with a narrow frequency range and a fixed power output (in the range from about 50 W to 4 kW). In the early days the piezoelectric material used was quartz, but more recently barium titanate (see Section 4.2) has become extremely important. The advantages of a synthetic piezoelectric material like barium titanate include flexibility in the choice of shape of transducer and (relative) cheapness of the transducer material; however barium titanate has the disadvantage that it cannot be used at temperatures above about 80°C. The materials most commonly used in ultrasonic cleaning processes using magnetostrictive transducers are nickel and certain ferrites.

Among the very many applications of ultrasonic cleaning in use nowadays are the cleaning of castings, wires and cables, old Roman coins,

Fig. 8.4. Automatic five-stage ultrasonic cleaning plant. (Photograph by courtesy of Dawe Instruments Ltd.)

cutlery, cine film, dies, moulds, and spinnerets, parts of internal combustion engines, dental and surgical instruments, and many other examples.

8.3. *Homogenizers and homogenization*

An emulsion consists of two liquids which are normally immiscible but which have been 'coerced' into forming a 'mixture' by dispersing very small particles of one liquid as a suspension in the second liquid. There are a number of influences acting on the suspended particles, some of which favour the separation of the two liquids while others act to oppose the separation. In many cases the latter influences are dominant and the emulsion is stable and can be stored 'on the shelf' for many years without separating. Some emulsions, e.g. milk, occur naturally. In the foodstuff, drugs, and cosmetic industries there are many emulsions which it is desirable to manufacture artificially. The traditional method of manufacture involved mechanical beating or churning of the ingredients to break up one of the liquids into small enough particles of the suspended liquid to form an emulsion.

Although we shall continue to use the word 'emulsion', we shall actually use the words 'homogenization' for the process of making an emulsion and 'homogenizer' for the machine that is used, instead of the words "emulsification" and "emulsifier", respectively. This is to avoid confusion with the earlier use of the word 'emulsifier' to describe a chemical which acts as an emulsifying agent to promote emulsification or homogenization.

It appears to have been first suggested by Wood and Loomis in 1927 that it might be possible to produce an emulsion by subjecting the interface between two immiscible liquids, such as oil and water, to ultrasound. At that time the implication was that the ultrasound would be generated by using quartz transducers. For many years not much progress was made towards developing a viable commercial ultrasonic homogenization system using either quartz piezoelectric transducers or magnetostrictive transducers. Then, perhaps somewhat surprisingly, it came to be realized that in this instance, unlike most other technological applications of ultrasound, the clue to developing a practical system was to be found in mechanical generators of ultrasound rather than in the use of transducers and electrical oscillators.

In the 1940s Janovsky and Pohlmann proposed the use of a 'liquid whistle' as the ultrasonic generator and, eventually, it was this which came to be adopted when commercial ultrasonic homogenizers became available. We are familiar enough with 'air whistles' in which a stream of air is directed into a resonant cavity or is made to impinge on some obstacle and in Section 4.1 we have, indeed, discussed the extension of the range of such whistles to ultrasonic frequencies. However, air is not the only fluid that can be used and one can use other gases, although this is of little relevance in the present context. Also, one can use a jet of liquid instead of a jet of gas. Whereas air whistles evolved over the centuries, liquid whistles have been developed quite recently and have been designed rather than allowed to evolve.

In principle, any type of air whistle could be modified to work with a liquid in place of air. But the type that is widely used commercially in homogenization processes is the 'jet-edge' generator which is based on the idea illustrated schematically in fig. 8.5. Vibrations may be set up either in the fluid in the region XY or in the edge itself; in the liquid

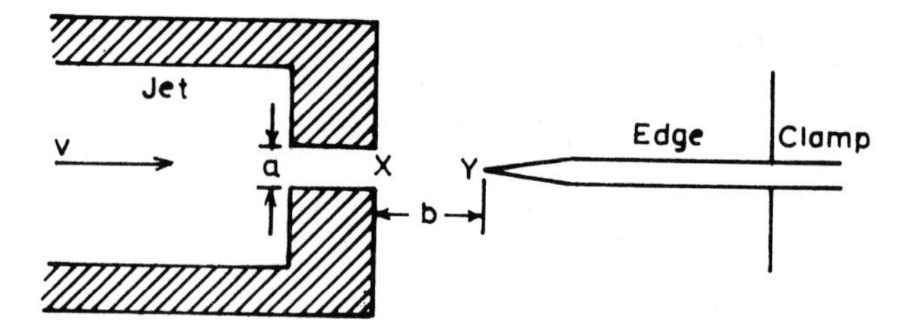

Fig. 8.5. Jet-edge system. (From B. Brown and J.E. Goodman, *High Intensity Ultrasonics* (Iliffe, London, 1965).)

whistle both types of vibrations are important. One can design a liquid whistle so that these vibrations occur at a common frequency. In an air whistle it would only be the former type of vibrations which would be important and the whistle would be designed with a suitable resonant cavity surrounding this region. For a straight edge it was shown by Lord Rayleigh that the frequency of sound, or ultrasound, generated in the fluid is given by

$$\nu \simeq \frac{nv}{b}\left(A + \frac{B}{R}\right) \tag{8.2}$$

where n (= 1, 2 or 3) characterizes a particular stage in the operation of the jet, A and B are constants, v is the velocity of the fluid through the jet, and R is Reynold's number for the fluid. However, in a liquid whistle the (mechanical) resonant vibrations of the edge itself are extremely important and so a great deal of work has gone into designing the edge. The edge consists of a steel blade, with a bevelled edge facing the jet. The plate is mounted in such a way that one particular normal mode of flexural vibrations will be excited. The plate may be mounted as a cantilever or at its half-wave nodal points. For a plate of thickness a, length l, density ρ, and Young modulus E, the frequency ν of the fundamental flexural vibrations is given by

$$\nu = \frac{\pi a}{2\sqrt{12}\,l^2}\sqrt{\left(\frac{E}{\rho}\right)} \times (0{\cdot}597)^2 \tag{8.3}$$

for cantilever mounting (see fig. 8.5) or by

$$\nu = \frac{\pi a}{2\sqrt{12}\,l^2}\sqrt{\left(\frac{E}{\rho}\right)} \times (2{\cdot}5)^2 \tag{8.4}$$

for the mounting shown in fig. 8.6. These formulae neglect the bevelling of the edges and the damping effect of the liquid. Overtones can also occur, but they appear not to be very important in practice.

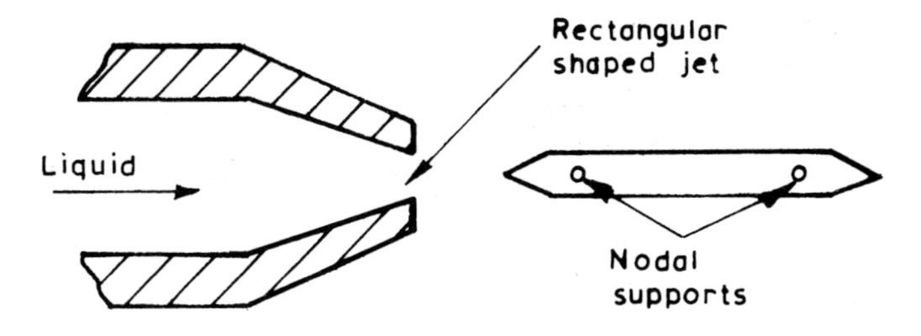

Fig. 8.6. Liquid whistle with nodally supported plate. (From B. Brown and J.E. Goodman, *High Intensity Ultrasonics* (Iliffe, London, 1965).)

In a machine for ultrasonic homogenization, the two liquids which are to be homogenized are forced together through the jet and the ultrasound is therefore either generated directly in the liquids as the edge tones or is very quickly transferred to the liquids from the vibrations of the blade. The energy requirement of a liquid-whistle homogenizer is very much lower–perhaps by a factor of about 10–than that of a comparable machine involving mechanical shaking or stirring of the component liquids. The ultrasonic homogenizer is therefore a rather compact unit, consisting of a pump and the whistle, and it can be installed 'on line' between the storage tanks for the ingredients and for the product. Following the successful development of commercial ultrasonic homogenizers, applications have been found for them in a wide range of industrial processes, principally in the manufacture of foodstuffs, drugs, and cosmetics, but in a number of others as well. The list of products that can be manufactured by ultrasonic homogenization is now very extensive. A few random examples from the foodstuffs applications are the manufacture of baby foods, condensed milk, fruit juices and pulps, gravies and soups, ice cream, ketchup, margarine, mayonnaise and salad cream, peanut butter, and processed cheese.

Another possibility which has been tried and which may become important in the future is to use an ultrasonic homogenizer to form emulsions of fuel oil and water (say about 3:1) to burn in boilers to increase their efficiency. A system due to Cottell was described in the journal *Ultrasonics* in March 1975 (see fig. 8.7). The emulsion, after it has been formed, is broken up into small particles of oil (about 0·01 mm in diameter) containing very much smaller particles of water in suspension. As combustion takes place the small particles of water turn into steam instantaneously and break up the particles of oil producing a large surface area for combustion. The efficiency improves because the water has replaced about 30 per cent of the air in the combustion process, eliminating

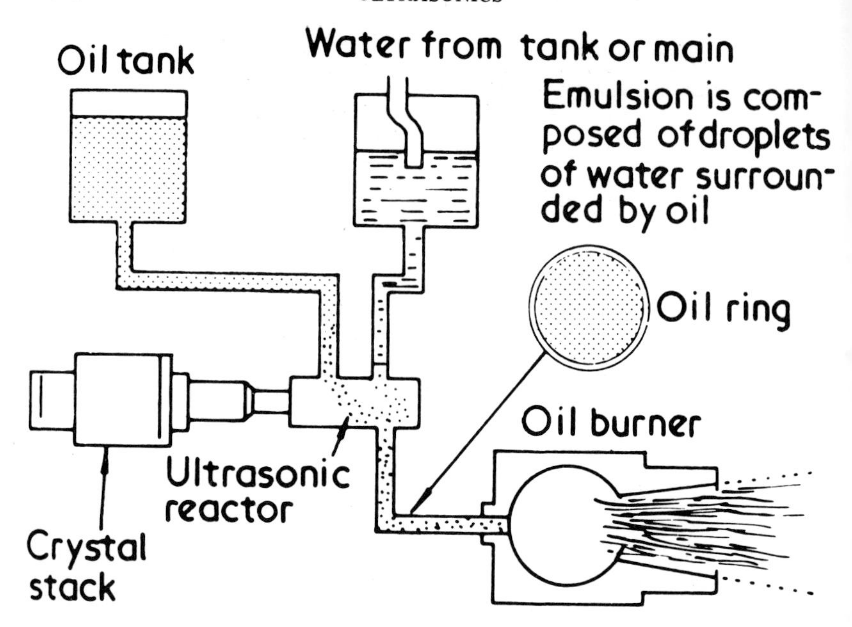

Fig. 8.7. Ultrasonic combustion system due to Cottell. (From *Ultrasonics*, 1975, **13**, 51.)

much of the nitrogen that consumes energy by dissociating. The heat transfer is also much more efficient because, by using less air, there is less nitrogen and excess oxygen in the combustion products. An added advantage is that the flame is cooler than in a conventional burner and so smaller quantities of pollutant oxides of nitrogen are produced. Another possibility that has been suggested is to use a 3 : 1 : 1 mixture of pulverized coal, oil and water which is said to require few changes of design to burn in conventional oil burners.

8.4. *Working of metals, plastics, etc.*

In some circumstances the use of ultrasound in connection with the working of metals, plastics, etc. involves the exploitation of the mechanical agitations of the ultrasonic vibrations. This may be a direct use of the vibrations, as in the case of an ultrasonic drill or saw. Thus, for example, in an ultrasonic drill the drill bit is coupled mechanically to a transducer in which longitudinal ultrasonic oscillations are being generated and thus the drill bit is made to vibrate longitudinally as well. This means that in an ultrasonic drill the action is reciprocal, like that of a pneumatic road drill or a hand saw, rather than rotary, like that of a hand drill or electric

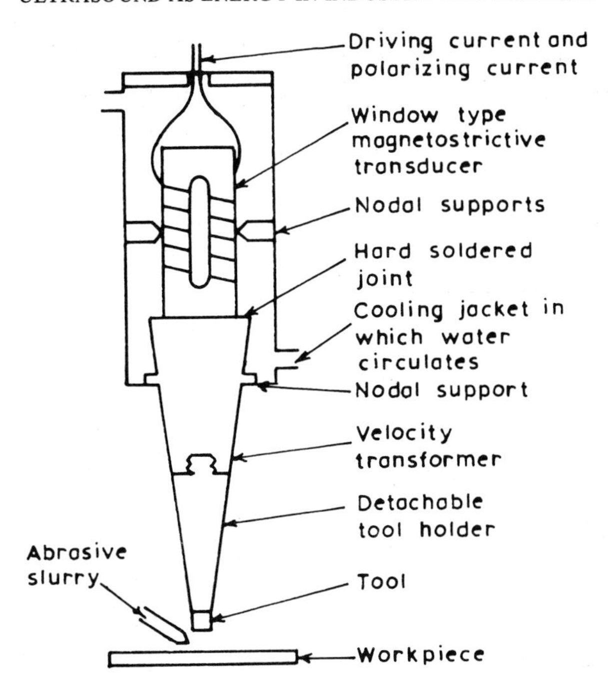

Fig. 8.8. Ultrasonic drill. (From B. Brown and J.E. Goodman, *High Intensity Ultrasonics* (Iliffe, London, 1965).)

drill. The cutting is performed by an abrasive slurry which is circulated continuously between the drill bit and the material which is being drilled. The drill bit is, therefore, not itself cutting the material directly but is thrusting the abrasive violently into contact with the material; consequently the drill bit need not be made of a specially hard material and may be made of mild steel. Ultrasonic drills are particularly suitable for work on hard and brittle materials and such machines are available commercially. An ultrasonic drill is illustrated schematically in fig. 8.8. One interesting feature of ultrasonic drilling is that it can be used to drill square, or other non-circular, holes in hard and brittle materials. This arises from the fact that the drilling action is reciprocal, instead of rotary, and the shape of the hole will be the same as the cross section of the bit. Holes of extremely complicated shape can also be drilled simply by making and using a bit with cross section of the required shape. As with most other drills, the bits are detachable and interchangeable. Although, in principle, one could use either piezoelectric or magnetostrictive transducers in an ultrasonic drill, for a number of practical reasons it is almost universal to use magnetostrictive transducers.

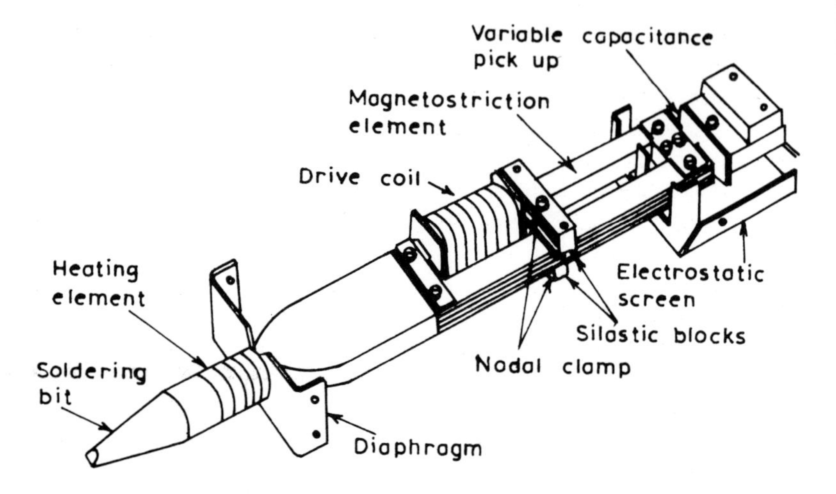

Fig. 8.9. Internal construction of Mullard ultrasonic soldering iron.

Another way in which the high frequency of ultrasonic vibrations can be exploited rather directly is in fatigue-testing of materials. The advantage of using the high frequency is that a test, which consists of a prescribed number of applications of a stress, can be completed in a much shorter time than is possible with more conventional methods. The high frequency of ultrasonic vibrations can also be exploited in the construction of bubble chambers (see Section 9.4).

Instead of exploiting the mechanical vibrations quite so directly as in the above examples, these vibrations may be used to act as a 'catalyst' to assist the conventional operation of a process. Thus in processes like soldering, tube-bending, and wire-drawing, ultrasound is applied to the job in addition to, and not as a replacement for, the conventional stress or source of energy used in the process. For example an ultrasonic soldering iron has been developed and manufactured (see fig. 8.9). Its design is clearly related to that of a conventional soldering iron. In the use of a conventional soldering iron, a flux is used to clean the surfaces of the metallic parts that are to be joined; this flux acts chemically to remove the oxide coatings that usually cover metallic surfaces. The idea behind the use of ultrasound is to replace the flux and to use the ultrasound, propagating in the molten solder, to clean the two surfaces that are to be joined. (The principles involved in ultrasonic cleaning have already been described in Section 8.2.)

One of the principal advantages of using ultrasound in soldering is in soldering aluminium. Aluminium has been used extensively in certain situations as an electrical conductor that is comparatively cheap, but there are many other situations in which previously it was not possible to replace copper by aluminium because of the difficulty of soldering it. One example is in the windings of many generators and motors. It is very difficult to clean aluminium using conventional fluxes, because the oxide Al_2O_3 is more resistant to attack by most typical fluxes than is the underlying metallic aluminium itself. The other conventional possibility, that of mechanical scrubbing of the aluminium surface beneath a protective covering to prevent the oxide from reforming, is also tedious and not very satisfactory. By using ultrasound it has become much more feasible to solder aluminium commercially and this has led to the replacement of copper by aluminium in, for example, the windings of certain motors. However, in practice, ultrasonic soldering is not usually performed with an ultrasonic soldering iron. Instead, dip-soldering is used in which ultrasound is propagated in a bath of molten solder and the two surfaces to be soldered are held together and dipped into the bath. As an alternative to joining two metals together, the ultrasonically agitated bath of molten solder can also be used for plating small parts and wires.

Yet another important field in which the mechanical vibrations of ultrasound can be exploited indirectly is as a source of localized heat in situations in which a direct application of heat by a flame, or an electrically heated metallic element, would be too destructive. Thus ultrasonic welding of plastics is quite frequently employed. The use of ultrasonic energy in the welding of thermoplastic components dates from the early 1960s. A diagram of a system which may be used for the purpose is given in fig. 8.10. The pieces that are to be welded together are held under pressure while ultrasonic vibrations are applied to the job from a transducer via the welding horn. The heat which is generated melts the plastic which flows under the applied pressure. When the vibrations are stopped the plastic solidifies again to form a strong weld. In addition to welding plastic to plastic, there is also the possibility of staking plastic to metal (fig. 8.11), or inserting metal into plastic (fig. 8.12). Ultrasonic welding of plastics, including staking and insertion, is now extremely widespread in the assembly of many different objects. A recent booklet from Dawe Instruments Ltd. listed about 40 examples, ranging from cotton reels, kitchen scales and door chimes, to car rear-lamp assemblies, cameras, radios, vacuum cleaners, and toys and remarked that 'the range is being added to daily'.

In addition to the use of ultrasound in connection with the welding of plastics, metals too can be welded ultrasonically. We have mentioned one advantage of using ultrasound in welding, namely that it avoids having an intense source of heat in the vicinity of the joint that is to

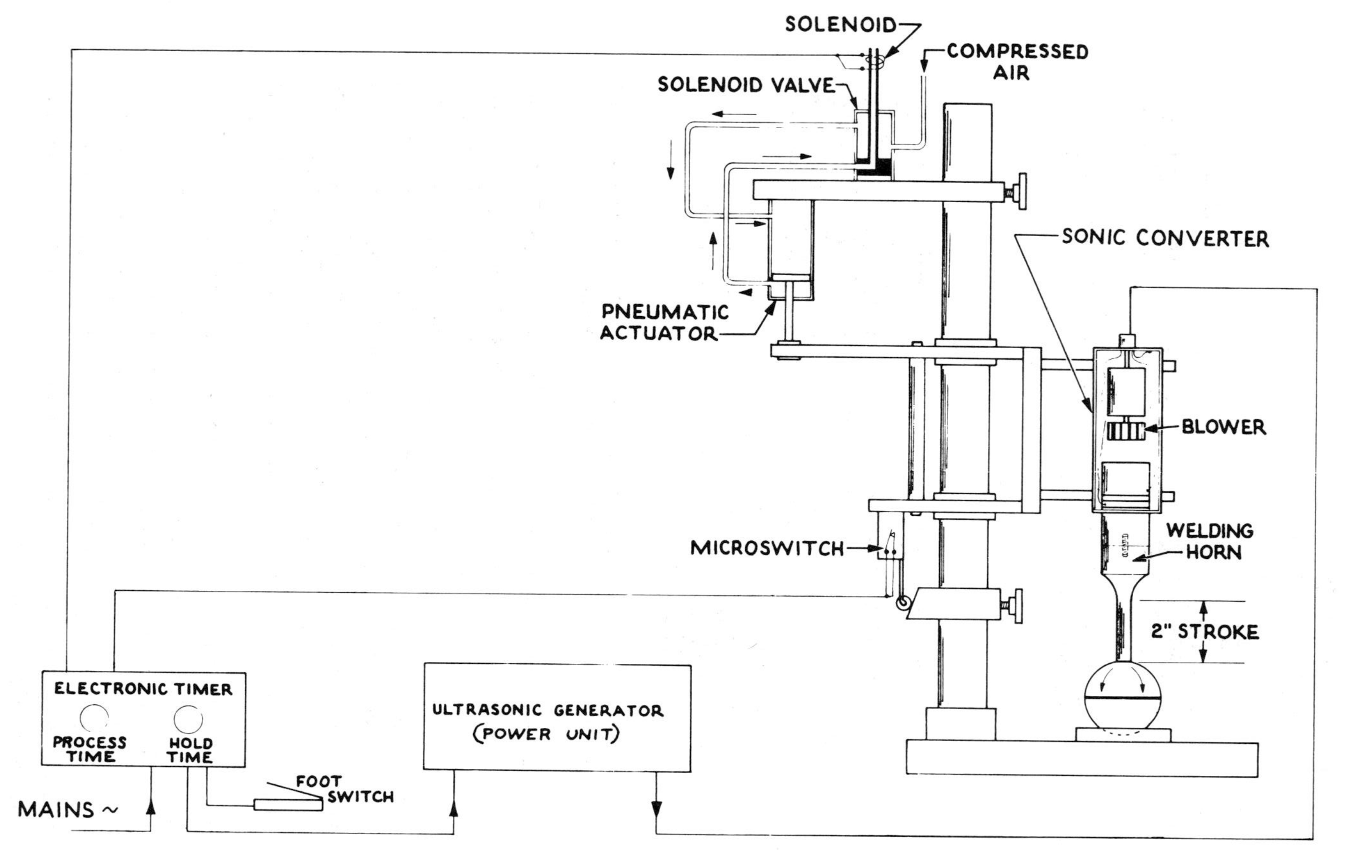

Fig. 8.10. Ultrasonic plastic welder. (By courtesy of Dawe Instruments Ltd.)

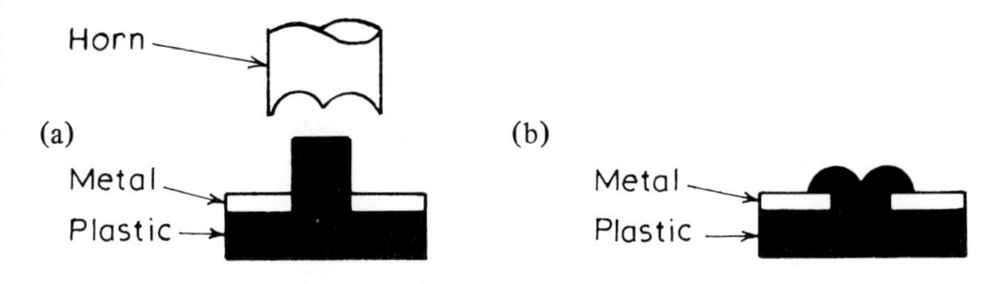

Fig. 8.11. Diagram to illustrate the staking of plastic to metal, (*a*) before and (*b*) after the application of ultrasound.

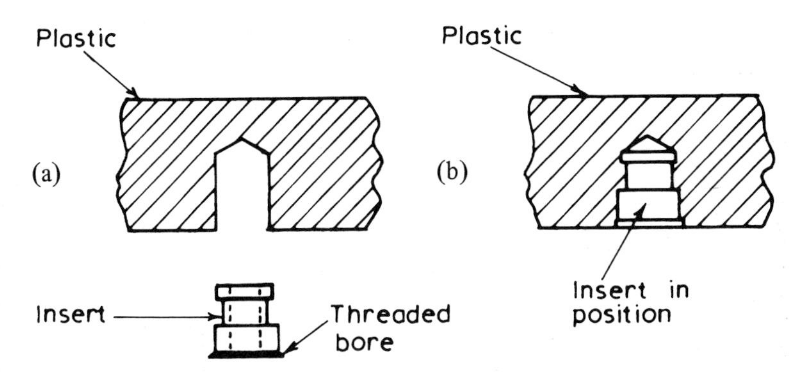

Fig. 8.12. Diagram to illustrate metal insertion into plastic (*a*) before and (*b*) after insertion and the application of ultrasound.

be welded; this causes much less heating of the surrounding parts of the job than would occur by the more conventional application of heat from a flame or a heated piece of metal. Naïvely, one might imagine that the physical mechanism involved in ultrasonic welding involves the conversion of the ultrasonic energy into the form of heat *in situ*. However, it now seems that, at least for metals, the welding does not occur by melting of the surfaces that are being welded together. What seems more likely is that the welding occurs as a result of ultrasonically-induced plastic deformation of the metal surfaces in contact. This is very similar to the seizure which is very likely to occur between two moving metallic surfaces which are in contact without any lubricant between them. Ultrasonically induced recrystallization of the surfaces is another possible, but less likely, mechanism for the welding. Apart from the avoidance of the use of an intense source of heat, the other advantage of using ultrasonic welding of metals include the fact that little surface preparation is necessary, only slight deformation of the metal occurs, and there should be no arc, sparks, or smoke produced.

8.5. *Ultrasonic surgery*

Probably the most important medical application of ultrasound at the present time is in various ultrasonic diagnostic techniques (see Section 5.3). A second important area of medical applications involves the use of ultrasonic cleaning baths for the cleaning of dental and surgical instruments (see Section 8.2). In dentistry, in addition to the cleaning of instruments by immersing them in a bath of cleaning fluid which is vibrated ultrasonically there is a variation on the use of ultrasonic cleaning. This involves using a specially designed ultrasonic probe, backed up by the appropriate commercially supplied instrumentation, for the scaling of teeth. A third important area of medical applications of ultrasound is in surgical work. The use of ultrasound in surgery was introduced much more recently than the diagnostic and cleaning techniques which we have already mentioned. Several of the possible surgical applications are still at the experimental stage and are not yet established as routine practice.

There are two distinct and alternative principles which may be employed in ultrasonic surgery. The first involves the application of a focussed beam of high-power ultrasound to small-sized and relatively inaccessible parts of the body. The purpose here is to induce changes by a very localized application of energy without damaging the surrounding regions. The frequencies used are of the order of MHz. The second possibility involves the use of surgical instruments which are vibrated ultrasonically. This is very similar to the use of ultrasonic agitation in cutting, drilling, and sawing of metals, rocks, etc., which we have mentioned in Section 8.4, and so the frequencies used for the ultrasonic agitation of surgical instruments are of the order of tens of kHz.

There have been many serious experimental investigations of the possibility of the use of focussed beams of ultrasound as a means of achieving a localized application of energy. This use of ultrasound has begun to become widespread in ear surgery and we shall outline the ultrasonic treatment of Ménière's disease. This disease is a disorder of the balance mechanism of the inner ear which causes sudden and disabling attacks of vertigo accompanied by nausea; it continues progressively to total deafness in the affected ear. The ultrasonic treatment of this disease has been performed by irradiating the surgically-exposed semi-circular canal with ultrasonic energy at 3 MHz (see fig. 8.13(*a*)). There are, however, disadvantages with this method. First, there is the time involved in the surgical removal of the overlying mastoid bone; this is a delicate operation because it involves the risk of total loss of hearing in that ear if the semi-circular canal is pierced. Secondly, the facial nerve passes close to the area to which the ultrasonic energy is applied and there is a danger that the heating effects may cause facial paralysis. An

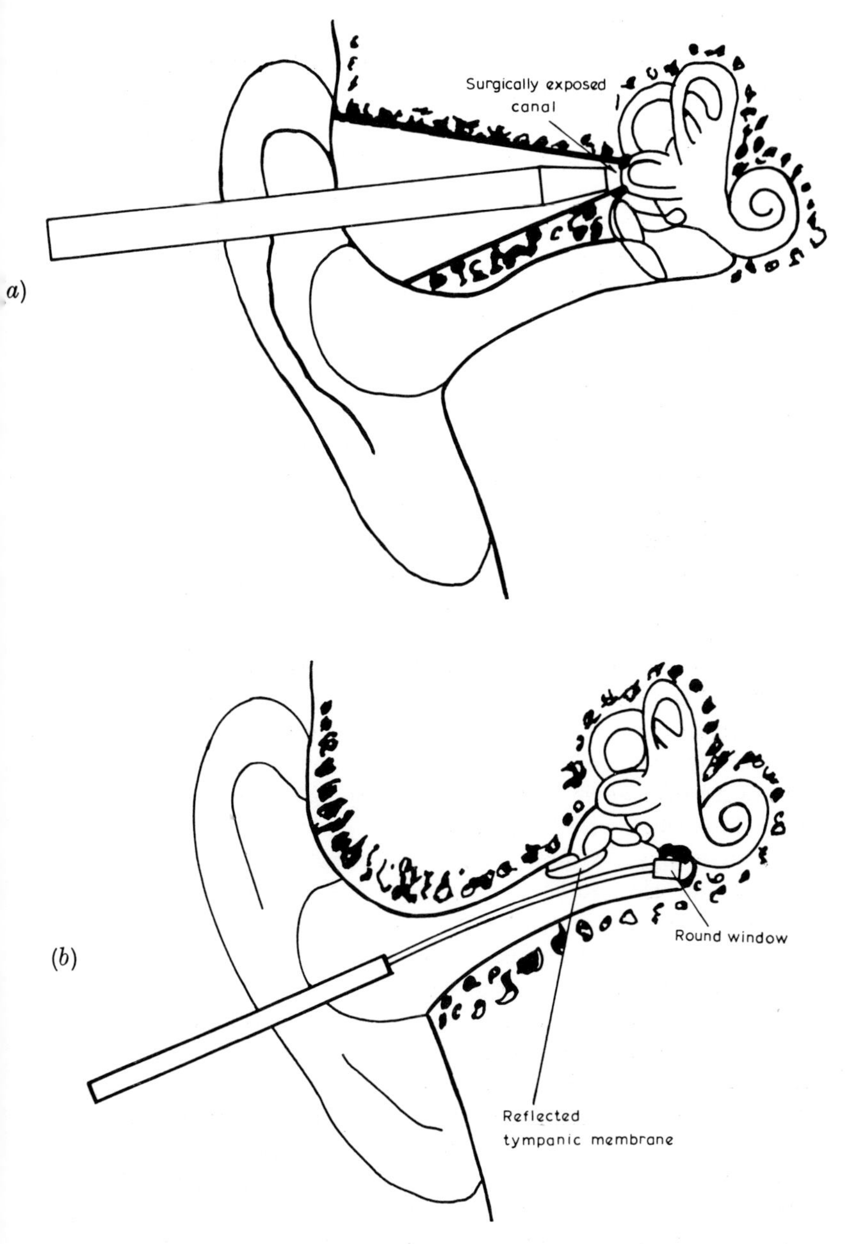

Fig. 8.13. Ultrasonic techniques for the treatment of Ménière's disease, (*a*) semi-circular canal technique and (*b*) round window technique. (From G. Kossoff, *Ultrasonics International 1973 Conference Proceedings*, p. 199, (IPC Science and Technology Press, Guildford, 1973).)

alternative treatment which eliminates these disadvantages involves placing a small ultrasonic applicator beside the round window after lifting aside the ear drum by surgery ('reflection of the tympanic membrane') (see fig. 8.13(*b*)). The success rate for the round window treatment is similar to that of the semi-circular canal treatment.

The possibility of the use of ultrasonic vibrations as a 'catalyst' for cutting, drilling, or 'welding' work in a medical context was first investigated in experiments on animals in the late 1960s. The cutting originally involved the use of an ultrasonic scalpel for cutting brain tissue, but this has now been extended to using ultrasonic knives and saws in the cutting of soft tissues and bones. In these cases the surgeon is not having to use a completely new technique but is operating with the conventional form of instrument, such as scalpel or saw, which has had a new 'catalyst' of ultrasonic vibrations added to it. The advantage of applying ultrasonic vibrations to a surgical instrument is that, as we have already noted in Section 8.4, this reduces very greatly the force that is needed for cutting. This, in turn, reduces the surgical trauma or shock suffered by the patient. In addition to reducing trauma, ultrasonic vibrations have some other useful effects when applied to surgical instruments. For example, the use of an ultrasonic knife may reduce bleeding in certain types of operation.

Whereas ultrasonically assisted cutting and sawing in surgery is closely analogous to the cutting and sawing of metals, etc., the surgical ultrasonic 'welding' of bones is less directly analogous to the welding of metal or plastics. The ultrasonic 'welding' of bones is more aptly described as osteosynthesis. When using ultrasound in osteosynthesis for knitting bone fragments, a liquid monomer, ciacrin, loaded with bone dust or bone shavings, is put into the gap. Ultrasound is then applied and a rigid joint is quickly formed. The ultrasound has two effects–one is to accelerate the diffusion of the liquid into the pores of the bone fragments that are to be joined, while the other is to accelerate the polymerization of the liquid to form a solid or hardened 'glue' holding the bone fragments together. It is interesting to note that it is thought that this 'glue' does not inhibit the natural bone regeneration and that, as this occurs, the 'glue' is slowly dissolved away.

In addition to the usual requirements for ultrasonic instruments for other purposes, an ultrasonic instrument for surgical work has to satisfy some special requirements. These include the minimization of the size and weight of the instrument, the resistance to corrosion, the suitability for sterilization, the safety of patient and surgeon, and the convenience and reliability of monitoring and control. It seems that magnetostrictive transducers are the most likely type of transducers to satisfy these requirements; they may be made of ferrites or, if the acoustic loading is heavy, a magnetostrictive metal may be used. Apart from the examples of the surgical uses of ultrasound which we have mentioned there are many others which have also been explored (see Further Reading).

9. Scientific applications of ultrasound

In these applications we shall consider the use of ultrasound to assist in physical investigations. This may be done in two ways. Ultrasound may be used as a probe to assist in studying some physical property of a body or system; most of the applications discussed in this chapter come into this category. Or ultrasound may be used in the construction of a scientific instrument, such as a bubble chamber.

9.1. *The determination of elastic moduli; propagation of ultrasound in crystals*

It will be recalled from Chapter 2 that the velocity of ultrasound is related to the elastic properties of the medium in which the ultrasound is propagating. There are many materials that are simply not available in large enough specimens of a convenient shape for use in the conventional stretching, bending, or squeezing experiments for determining the elastic constants. The measurement of the ultrasonic velocities then provides a convenient method for the determination of the elastic constants of such materials. Moreover for many materials, although it is possible to perform conventional stretching, bending, or squeezing experiments, it is a good deal easier to make ultrasonic velocity measurements (see Section 7.1). This is particularly true for gaseous or liquid specimens for which it is much less difficult to establish good acoustic contact with the specimen than is the case with solid specimens.

The elastic properties of an isotropic solid can be fully described by the three elastic moduli: the Young modulus E, the bulk modulus K, and the shear modulus G We have seen in Section 2.1 that the velocity of ultrasound in an isotropic medium is given by

$$c = \sqrt{\left(\frac{\mathscr{E}}{\rho}\right)} \tag{2.22}$$

where $\mathscr{E}$ is the appropriate modulus, or combination of moduli, of elasticity and ρ is the density. In the context of the present chapter, the purpose of such experiments can be regarded as the determination of moduli of elasticity. The expressions for $\mathscr{E}$ for various important types of waves were summarized in terms of E, G, and K on page 17. For a gas, the determination of the bulk modulus, by ultrasonic means or otherwise, can be regarded as a way of determining γ, the ratio of the two principal specific heat capacities c_p/c_v.

The elastic properties of an isotropic material can be fully described by, at the most, three moduli of elasticity (with the possible addition of Poisson's ratio), but the elastic properties of an anisotropic crystalline material are more complicated. Crystals are anisotropic as far as the propagation of ultrasound is concerned, that is the velocity of ultrasound will not be the same for all possible different directions of propagation relative to the crystal axes.* This statement should be taken to apply to single-crystal specimens. For a specimen which is polycrystalline, the average effect of having a large number of tiny crystals in random orientations, relative to one another, will be to give a material that is isotropic in its general physical properties and, in particular, for the propagation of ultrasound.

In a crystal both the stress and the strain have to be represented by tensors of rank two. We define the stress tensor T_{ij} as follows: T_{ij} is the force component in the direction parallel to the x_i axis ($(x_1, x_2, x_3) = (x, y, z)$) and acting on the face that is normal to the x_j axis; some of these components are illustrated in fig. 9.1. Thus, for example, the force T_{11} acts in the x direction on the face ABCD in fig. 9.1 which is normal to the x axis. There will be an equal and opposite force on the face EFGH at the back and these two forces together will provide a tension that stretches the specimen in the x direction. If T_{11} has a negative value it will correspond to a compression instead of a stretching of the specimen. T_{12} is also a force in the x direction but it is acting tangentially on the face CBFG; together with the corresponding equal and opposite force on the face DAEH this provides a shear stress on the specimen. We can define a strain tensor too. Suppose that P and Q are two points with coordinates x_i and $(x_i + \delta x_i)$ respectively in an unstrained crystal. Suppose also that the displacements of these points in the strained state are u_i and $(u_i + \delta u_i)$ respectively. It is possible to show that δu_i, the difference between the displacements of Q and P can be regarded as a rotation of the specimen combined with a local deformation. It is possible to show that the part of δu_i corresponding to a deformation can be written in the form

*Crystals with cubic symmetry, such as rock salt, diamond, or even iron, may be regarded as isotropic as far as some of their physical properties, such as the electrical conductivity, are concerned. However, this is not the case for the elastic properties or, therefore, for the propagation of ultrasound.

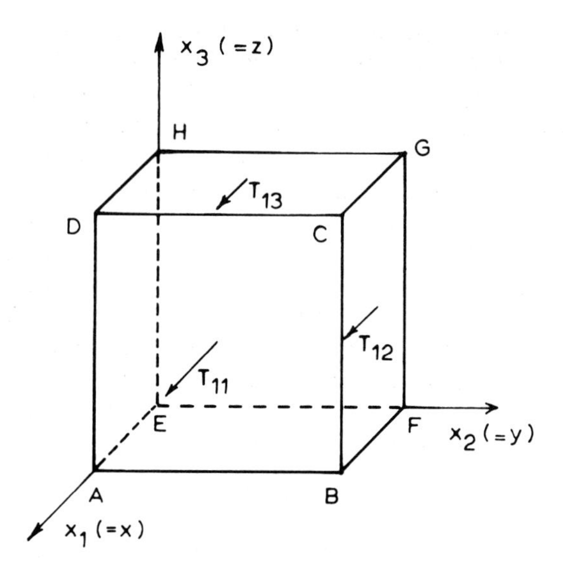

Fig. 9.1. Diagram to illustrate the definition of stress in a crystal.

$$\sum_k \eta_{ik} \delta x_k = \sum_k \frac{1}{2}\left(\frac{\partial u_i}{\partial x_k} + \frac{\partial u_k}{\partial x_i}\right) \delta x_k \tag{9.1}$$

and the quantity

$$\eta_{ik} = \frac{1}{2}\left(\frac{\partial u_i}{\partial x_k} + \frac{\partial u_k}{\partial x_i}\right) \tag{9.2}$$

is called the *strain tensor.*

Since the stress and the strain in a crystal are both described by tensors of rank two, the generalized version of Hooke's law which replaces equations like eqns. (2.3), (2.6) and (2.9) for a crystal is

$$T_{ij} = \sum_{k,l} C_{ijkl} \eta_{kl} \tag{9.3}$$

where each of the suffices i, j, k, and l may take the possible values 1, 2, and 3. C_{ijkl} is a tensor of rank four and the components of C_{ijkl} are called the *elastic stiffness constants*. As an alternative to eqn. (9.3) it is possible to write the components of the strain tensor in terms of the components of the stress tensor; thus

$$\eta_{ij} = \sum_{k,l} S_{ijkl} T_{kl} \tag{9.4}$$

where S_{ijkl} is also a tensor of rank four and the components of this tensor are called the *elastic compliance moduli*. Thus, instead of describing the elastic properties by just three numbers, E, K, and G, which apply to an isotropic solid, we have to describe the elastic properties of a crystal by specifying the values of all of the components of one of the fourth-rank tensors C_{ijkl} or S_{ijkl}. Since each of the suffices i, j, k, and l can take any value from 1 to 3 this means that there are $3^4 = 81$ components of the tensor C_{ijkl} or of the tensor S_{ijkl} which have to be specified. Fortunately, there are some simplifying relationships which can be used and which mean that we do not have to carry out experiments to measure 81 different numbers for each different crystalline material. Even for a material with the lowest possible crystallographic symmetry, namely that of the triclinic point group 1, not all these 81 components are independent. There are a number of intrinsic symmetries

$$C_{ijkl} = C_{ijlk} = C_{jikl} = C_{klij} \tag{9.5}$$

and

$$S_{ijkl} = S_{ijlk} = S_{jikl} = S_{klij} \tag{9.6}$$

These conditions reduce the number of independent tensor components from 81 to 21 for C_{ijkl} and for S_{ijkl}. For a crystal with higher symmetry than triclinic it is possible to make some further reductions in the number of independent tensor components by exploiting the crystallographic symmetry. The greatest possible reduction that can be achieved is for a cubic crystal when the number of independent components is reduced to three. These may be taken to be C_{1111}, C_{1122} and C_{2323}. Thus there are only three independent numbers that have to be determined experimentally in order to specify the elastic properties of a cubic crystal. However, it is important to realize that this does not mean that these three tensor components could be replaced by the moduli E, K, and G. Cubic crystals are anisotropic in their elastic properties; consequently they are also anisotropic for the propagation of ultrasound. That is, the velocity of ultrasound is different for different directions of propagation. This sometimes leads to some confusion because a cubic crystal is isotropic in its dielectric properties; therefore it is isotropic for the propagation of light, that is the velocity of light in a cubic crystal is independent of the direction of propagation. Of course, crystals of lower symmetry than cubic crystals are anisotropic as far as the propagation of both ultrasound and light are concerned.

The differential equation governing the propagation of elastic waves in an isotropic medium was given previously in eqn. (2.16). If one wishes to describe the propagation of elastic waves in a crystalline medium it is necessary to generalize eqn. (2.16). This can be done by applying Newton's second law of motion to a small volume element of the crystal. One has to obtain expressions for the resultant force in, say, the x_i direction and equate this to (mass × acceleration) for this direction. There will be a contribution to the resultant force in this direction from each of the six faces; two from T_{ii} tensile or compressional forces and four from T_{ij} shearing forces. After a certain amount of manipulation it is possible to show that the equation of wave motion becomes

$$\sum_{j,k,l} C_{ijkl} \frac{\partial^2 u_k(\mathbf{r}, t)}{\partial x_j \partial x_l} = \rho \frac{\partial^2 u_i(\mathbf{r}, t)}{\partial t^2} \tag{9.7}$$

where $u_i(\mathbf{r}, t)$ $(i = 1, 2, 3)$ are the components of a vector representing the displacement at position $\mathbf{r}$ in the crystal at time t. Equation (9.7) is the equation of wave motion for travelling or standing elastic waves in a three-dimensional anisotropic medium. It is the analogue, for three dimensions, of eqn. (2.16) which applied to waves in a one-dimensional system. The summations on the left-hand side of eqn. (9.7) mean that we have a set of complicated simultaneous differential equations for the three components $u_i(\mathbf{r}, t)$ $(i = 1, 2, 3)$. By taking for $u_i(\mathbf{r}, t)$ an expression describing a harmonic wave and substituting into eqn. (9.7), one can obtain an expression for the velocity of the wave in terms of the elastic stiffness constants–although the actual details of the analysis are quite complicated.

It is convenient to reduce the number of suffices in the elasticity tensor from 4 to 2 in the following manner. Since both T_{ij} and η_{ij} are symmetric tensors it is common to use a single suffix, running from 1 to 6, instead of two suffices:

$$\begin{array}{cccccc} T_{11} & T_{22} & T_{33} & \underbrace{T_{23}, T_{32}} & \underbrace{T_{31}, T_{13}} & \underbrace{T_{12}, T_{21}} \\ T_1 & T_2 & T_3 & T_4 & T_5 & T_6 \end{array}$$

and similarly for η_{ij}. The same thing can be done for C_{ijkl}:

$$\begin{array}{ccccc} C_{1111} & C_{1112} & C_{1113} & C_{1122} & \cdots\cdots \\ c_{11} & c_{16} & c_{15} & c_{12} & \cdots\cdots \end{array}$$

Then eqn. (9.3) can be written in the form

$$T_i = \sum_k c_{ik} \eta_k \tag{9.8}$$

where $i, k = 1, 2, \ldots, 6$. It is important to remember that this is only a shorthand notation so that, in spite of their appearances, T_i, η_k, and c_{ik}

are still tensors of rank two, two, and four, respectively. In this notation the non-zero independent elastic stiffness constants C_{1111}, C_{1122}, and C_{2323} for a cubic crystal will be relabelled as c_{11}, c_{12}, and c_{44} respectively.

The expressions for the velocity of ultrasound in a cubic crystal in terms of c_{11}, c_{12}, and c_{44} are quite complicated and depend on the direction of propagation relative to the crystal axes. We simply quote the results for one particular direction. For a wave travelling with **k** along [110] in a cubic crystal, the possible velocities are found to be

$$\begin{aligned} v_1 &= \sqrt{[(c_{11}+c_{12}+2c_{44})/2\rho]} \quad &[110] \\ v_2 &= \sqrt{[(c_{11}-c_{12})/2\rho]} \quad &[1\bar{1}0] \\ v_3 &= \sqrt{(c_{44}/\rho)} \quad &[001]. \end{aligned} \tag{9.9}$$

The zone axis symbols indicate the directions of the polarization vectors of the waves; thus the first wave is longitudinal, while the other two are transverse. Thus, for example, for germanium measured values of the velocities obtained by McSkimin in 1953 are

$$\begin{aligned} v_1 &= 5400\,\mathrm{m\,s^{-1}} \\ v_2 &= 2750\,\mathrm{m\,s^{-1}} \\ v_3 &= 3550\,\mathrm{m\,s^{-1}}. \end{aligned} \tag{9.10}$$

The details of the expressions in eqns. (9.9) for the velocities of the three waves for this particular direction of propagation are not very important for our present purposes; these details will, of course, be different for other directions of propagation. There are three important points to notice now. First, by measuring the velocities of propagation of elastic waves, probably using pulse-echo techniques (see Section 7.1), and using equations like eqns. (9.9) we have a method for determining the elastic constants c_{ij} of a cubic crystalline material. For germanium, for instance, one obtains from the above values of v_1, v_2, and v_3

$$\begin{aligned} c_{11} &= 1{\cdot}3 \times 10^{11}\,\mathrm{N\,m^{-2}} \\ c_{12} &= 4{\cdot}9 \times 10^{10}\,\mathrm{N\,m^{-2}} \\ c_{44} &= 6{\cdot}9 \times 10^{10}\,\mathrm{N\,m^{-2}}. \end{aligned} \tag{9.11}$$

Secondly, as in the isotropic case, the values of v_1, v_2, and v_3 are independent of the frequency, or of the wavelength, of the elastic waves and so the dispersion relations for the waves are linear (see fig. 2.6). Thirdly, the procedure can, obviously, be extended to the determination of the constants C_{ijkl} for non-cubic crystals.

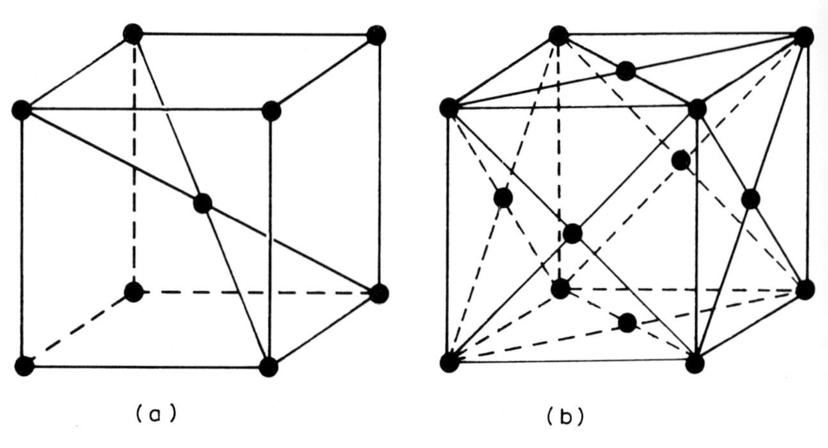

Fig. 9.2. Conventional unit cells of (*a*) body-centred cubic and (*b*) face-centred cubic structures.

9.2. *Phase transitions*

In Section 9.1 we discussed the relevance of ultrasonic velocity measurements to the determination of the elastic moduli or elastic stiffness constants of a material. The values of the elastic moduli or the elastic stiffness 'constants' of a solid may alter if the temperature changes and so the measurements of the ultrasonic velocities as functions of temperature enable the temperature dependences of the elastic constants to be studied. As a general rule the values of the elastic constants of a material change steadily as the temperature is varied. However, if a crystal undergoes a phase change there are likely to be quite substantial changes in the elastic constants, and therefore in the ultrasonic velocities, near the transition temperature. Consequently, by looking for abnormally large changes in the ultrasonic velocity, one has a possible method for detecting phase transitions.

Let us distinguish three different kinds of phase transition. First, there are some phase changes, such as the melting of a solid or the boiling of a liquid, that can easily be detected by the casual observer and do not need elaborate ultrasonic investigations to detect them. The second and third kinds of phase transition to which we refer are transitions between two different phases of a solid crystalline material. In the second kind of transition the two phases of the given material may have quite different structures. For example, iron has a body-centred-cubic structure below a certain transition temperature (909°C) but has a face-centred-cubic structure above this temperature (see fig. 9.2). The transition from one of these structures to the other involves a rather drastic rearrangement

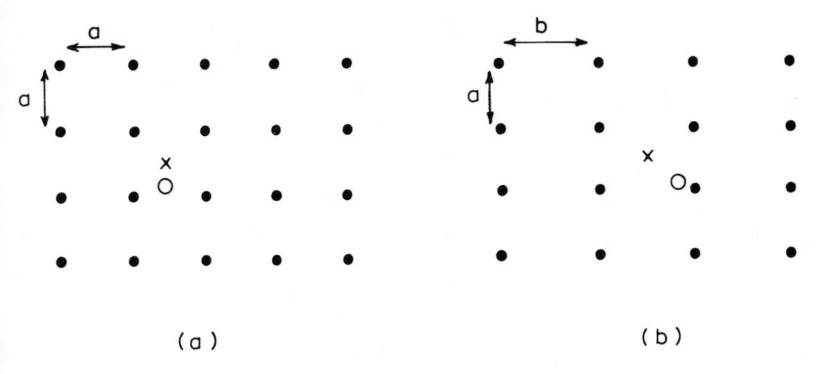

Fig. 9.3. A displacive phase transition.

of the positions of the atoms in the crystal relative to one another. A structural rearrangement of this type can be detected by using X-ray diffraction methods because the two structures give quite different X-ray diffraction patterns. Therefore there is no need to use ultrasonic measurements to detect the existence of one of these phase transitions. The third kind of transition is again between two different solid phases of the same material, but this time no drastic structural rearrangement is involved. This transition may be a displacive phase transition, or a magnetic phase transition, or a ferroelectric phase transition.

In a displacive transition displacements of the relative positions of the atoms in the structure occur; these displacements are only small but they nevertheless alter the qualitative features of the symmetry of the structure. This is illustrated schematically in fig. 9.3. In fig. 9.3(a) the dots, which represent atoms, are arranged in a square array, with the length of the edge of a unit cell denoted by a. In fig. 9.3(b) the array of dots has been distorted slightly so that it is rectangular instead of square. Let us suppose that this distortion has involved increasing the length of one pair of parallel edges of each unit cell to b instead of a, while leaving the other pair of parallel edges unaltered. The square lattice has more symmetry than the rectangular lattice. Thus an axis normal to the plane of the diagram and passing through the point O in fig. 9.3(a) is a four-fold axis of symmetry, that is the array of dots becomes indistinguishable from its starting position after it has been rotated by any integral number of quarter revolutions about this axis. But an axis normal to the plane of the diagram and passing through O in fig. 9.3(b) is only a two-fold axis of symmetry, that is the array of dots only reproduces its starting appearance after being rotated by any integral number of half revolutions. This difference between four-fold and two-fold rotational symmetry is a difference

Fig. 9.4. A ferromagnetic-paramagnetic phase transition (*a*) ferromagnetic material, $T < T_c$ and (*b*) the non-magnetic (paramagnetic) material, $T > T_c$.

of quality or of nature and the change from one to the other happens suddenly. As soon as *b* becomes at all different from *a*, no matter how small the actual magnitude of the difference is, the order of this axis of symmetry changes abruptly from four to two.

A second type of solid-solid phase transition that may not involve any drastic structural rearrangement is one in which magnetic ordering appears or disappears. Consider a material such as iron which is ferromagnetic. The individual magnetic moments in the material can be assumed to be aligned parallel to one another, thereby giving the material a net magnetic moment. If the temperature of the material is raised there is a certain temperature, known as the Curie temperature T_c, at which the spontaneous magnetic moment of the material disappears. This is illustrated schematically in fig. 9.4 in which it is assumed that at the Curie temperature it is only the orientations of the individual magnetic moments and not the relative positions of the atoms in the structure, which are altered. (This is not quite true because of magnetostriction, which involves a (rather small) change in the dimensions, and therefore in the unit-cell dimensions, when magnetization or de-magnetization occurs.)

A third important class of solid-solid phase transition that does not involve a drastic rearrangement of the atomic positions is that of ferroelectric phase transitions. A ferroelectric material is one which possesses a spontaneous electric dipole moment which is analogous to the permanent magnetic moment of a ferromagnetic material. One can imagine this to be due to a net spontaneous displacement of all the positive charges in the material, relative to all the negative charges. If the temperature of a ferroelectric material is raised steadily a transition temperature is eventually reached at which the ferroelectricity disappears. The appearance, and disappearance, of ferroelectricity as a material is cooled or heated through this transition temperature occurs, in many materials at least, without any drastic structural rearrangement of the atoms in the crystal.

Many examples can be found of materials that exhibit one or other of these three types of solid–solid phase transition that need not involve drastic structural rearrangement, namely displacive, magnetic, and

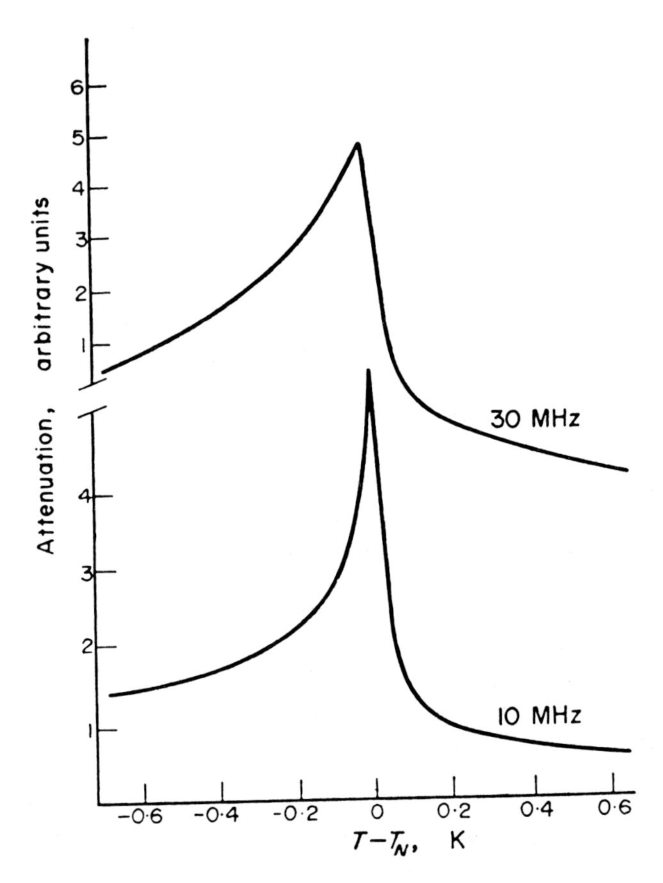

Fig. 9.5. Attenuation of 10 MHz and 30 MHz longitudinal ultrasound in MnF_2 in the [001] direction as a function of temperature near the Néel temperature. (From M.F. Cracknell and A.G. Semmens, 1971, *J. Phys. C (Solid State Phys.)*, **4**, 1513.)

ferroelectric transitions. It is extremely difficult, though not necessarily impossible, to detect these transitions by using X-ray diffraction; the changes that occur in the ultrasonic velocity may therefore provide a valuable means for detecting such transitions. Alternatively, and perhaps more commonly, the changes in the ultrasonic attenuation may be used in detecting these transitions or in determining the actual value of the transition temperature. Figure 9.5 illustrates the variation of the ultrasonic attenuation in MnF_2 as a function of temperature in the vicinity of 70 K. At about this temperature MnF_2 has a solid–solid phase transition that involves no drastic structural rearrangement. Below this temperature

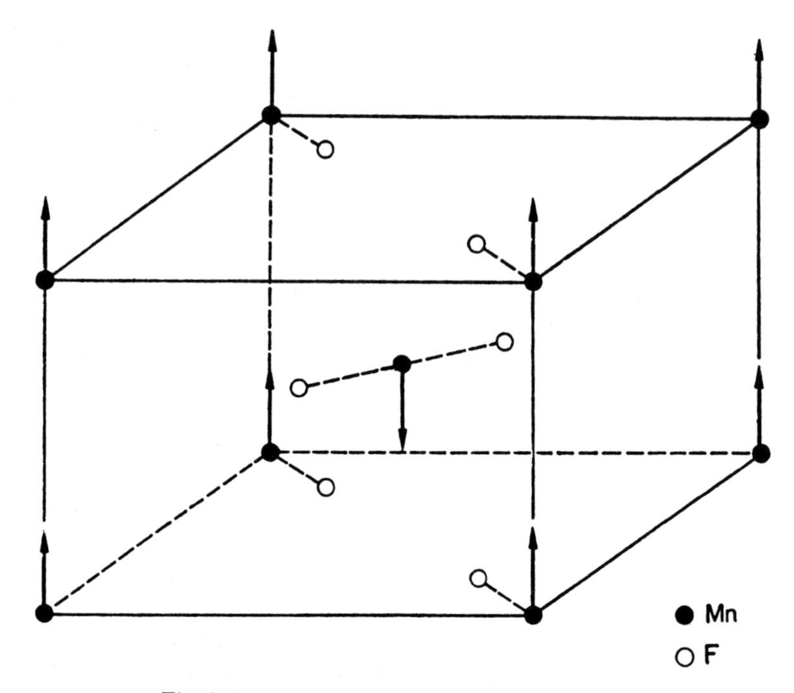

Fig. 9.6. Unit cell of antiferromagnetic MnF_2.

it is antiferromagnetic and above this temperature it is paramagnetic. The antiferromagnetic structure of MnF_2 is illustrated in fig. 9.6. This differs from that of a ferromagnet, such as iron, cobalt or nickel, in that there are equal numbers of magnetic moments parallel and antiparallel to a certain direction. Therefore, although there is a regular pattern of magnetic moments in an antiferromagnetic material, this material possesses no net magnetization. Consequently, it does not attract iron filings or deflect compass needles in the way that a ferromagnetic material would. In the paramagnetic phase the magnetic moments of the Mn(II) ions are randomly oriented.

9.3. *Electrons in metals and semiconductors*

The most interesting features of the behaviour of ultrasound in metals and semiconductors arise from the interaction between the ultrasound and the carriers of electric charge, namely electrons in metals and electrons and holes in semiconductors. We shall consider three topics (*a*) the attenuation of ultrasound in superconducting metals; (*b*) the attenuation of ultrasound (as a function of magnetic field) in normal, i.e. non-superconducting, metals; and (*c*) the acoustoelectric effect in semiconductors.

(*a*) *The attenuation of ultrasound in superconducting metals*. For any metal that exhibits superconductivity there is a certain temperature T_c which is called the *critical temperature*, or *superconducting transition temperature*; the value of T_c is characteristic of the metal. Below T_c the electrical resistance vanishes. Many, but by no means all, of the metallic elements exhibit superconductivity and it is also exhibited by many compounds and alloys. In addition to the vanishing of the resistance there are a number of other curious features of the properties of a superconducting metal below T_c. For example, the magnetic induction **B** within a superconducting metal is zero. Also, at a temperature below T_c the superconductivity of a specimen may be destroyed by a sufficiently large external magnetic field.

The states of a metal above and below T_c can be regarded as different phases, the *normal* and *superconducting* phases, of the metal. The phase transition at T_c occurs without any drastic rearrangement of the atomic, or ionic, positions within the metal. Therefore, in view of what was said in Section 9.2, it is reasonable to suppose that measurements of the ultrasonic velocity, or of the ultrasonic attenuation, as a function of temperature would provide a useful way of investigating the superconducting phase transition. Experiments performed on lead and on tin established that α_s, the attenuation of ultrasound in a superconductor, decreases rapidly as the temperature is lowered. If, at a temperature below T_c, the superconductivity is destroyed by the application of a sufficiently large magnetic field the attenuation returns to a higher value α_n, characteristic of the normal phase of the metal. This might not, at first sight, seem to be a particularly important or fundamental, but in fact the ultrasonic attenuation is very closely related to the underlying mechanism that is responsible for the existence of superconductivity.

In a superconductor there is an energy gap in the electronic density of states, due to the pairing of electrons to form *Cooper pairs*. A Cooper pair consists of two electrons with opposite spins and with equal and opposite momenta. If the temperature is sufficiently low the thermal energies of the electrons in the paired state are very much smaller than the gap energy; consequently these pairs transport charge through the metal without undergoing any collisions and therefore they suffer no electrical resistance. Cooper pairs are formed as a result of electron-phonon interactions. The phonons involved in the pair formation are the phonons of the thermal vibrations of the atoms (or ions) of the metal, but these phonons are not essentially different from the phonons of an ultrasonic wave. Consequently, it is not surprising that the attenuation of ultrasound in a metal should be profoundly affected by the onset of superconductivity. Moreover, it is also reasonable to expect that the

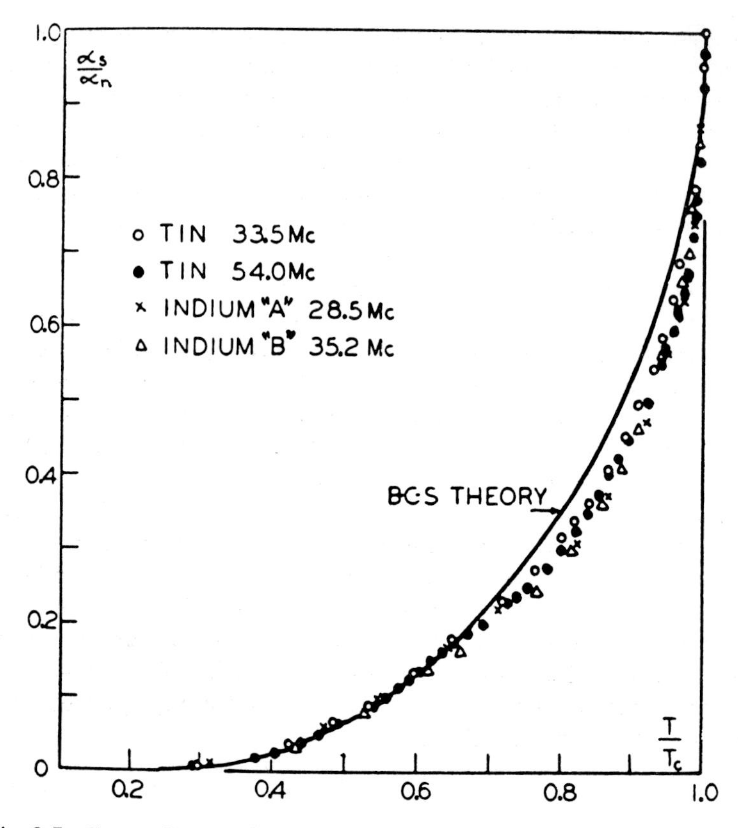

Fig. 9.7. Comparison between measured values of α_s/α_n for tin and indium and values calculated from the BCS theory (From R.W. Morse and H.V. Bohm, 1957, *Phys. Rev.* **108**, 1094.)

theory of Bardeen, Cooper, and Schrieffer (the BCS theory), which gives a satisfactory account of superconductivity from a microscopic point of view in terms of electron–phonon interactions, should also provide a satisfactory description of the ultrasonic attenuation in a superconducting metal. For an isotropic superconductor, the BCS theory predicts the ratio of the ultrasonic attenuation in the superconducting and normal states to depend on the energy gap as

$$\frac{\alpha_s}{\alpha_n} = \frac{2}{\exp(2\epsilon(T)/kT) + 1} \tag{9.11}$$

where $2\epsilon(T)$ is the gap, T is the temperature, and k is the Boltzmann constant. A comparison between the predictions of the BCS theory and experimental results for indium and tin is illustrated in fig. 9.7. Equation

(9.11) shows that it should be possible to use measurements of α_s/α_n to determine the superconducting energy gap. Indeed this is the case and such measurements have provided a very important method for measuring the gap. Moreover, by performing the experiments for different orientations of a single crystal specimen of a superconductor, it is possible to determine the anisotropy of the energy gap, that is the variation of the energy gap as a function of direction in a single crystal specimen of the metal.

(*b*) *The attenuation of ultrasound, as a function of magnetic field, in normal metals.* Under ordinary circumstances the conduction electrons in a metal are not localized within the individual atoms of which the metal is constituted; instead they are able to move around within the metal and they are free to move in different directions and with different speeds. If a magnetic field is applied to the metal, the paths of the conduction electrons will be altered. To help in understanding the effect of the magnetic field it is useful to consider the effect of a magnetic field on an electron in free space. The path of an electron moving in free space in a uniform magnetic flux density $\boldsymbol{B}$ is a helix, and the projection of this trajectory onto a plane normal to $\boldsymbol{B}$ is a circle. For convenience we can neglect the component of the motion parallel to $\boldsymbol{B}$ and loosely describe the orbit as a circle. The force that causes the circular motion is the Lorentz force $e\mathbf{v} \wedge \boldsymbol{B}$ on the moving electron so that, neglecting the component of the motion parallel to $\boldsymbol{B}$,

$$evB = mv^2/r. \tag{9.12}$$

Rearranging this equation and writing $v/r = \omega_c$ we obtain

$$\omega_c = eB/m \tag{9.13}$$

where $B = |\boldsymbol{B}|$. The frequency $\omega_c/2\pi$, which is independent of the radius of the orbit, is called the *cyclotron frequency*. Rearranging eqn. (9.12) gives

$$r = mv/eB. \tag{9.14}$$

Therefore, for a given value of the velocity the radius r of the circular orbit will be proportional to $1/B$. In a metal a conduction electron is by no means free since it is acted on by all the metal ions and by all the other conduction electrons in the metal; consequently we should expect the orbit of a conduction electron to be quite distorted from the shape of the orbit of a free electron. The orbit, however, is still a helix with its axis parallel to the direction of the magnetic flux density $\boldsymbol{B}$ and the projection of the orbit onto a plane normal to $\boldsymbol{B}$ will still be a closed orbit (see fig. 9.8). It is still possible to use an equation of the form of eqn. (9.13) where the free-electron mass m is replaced by an effective mass, m_c^*, which

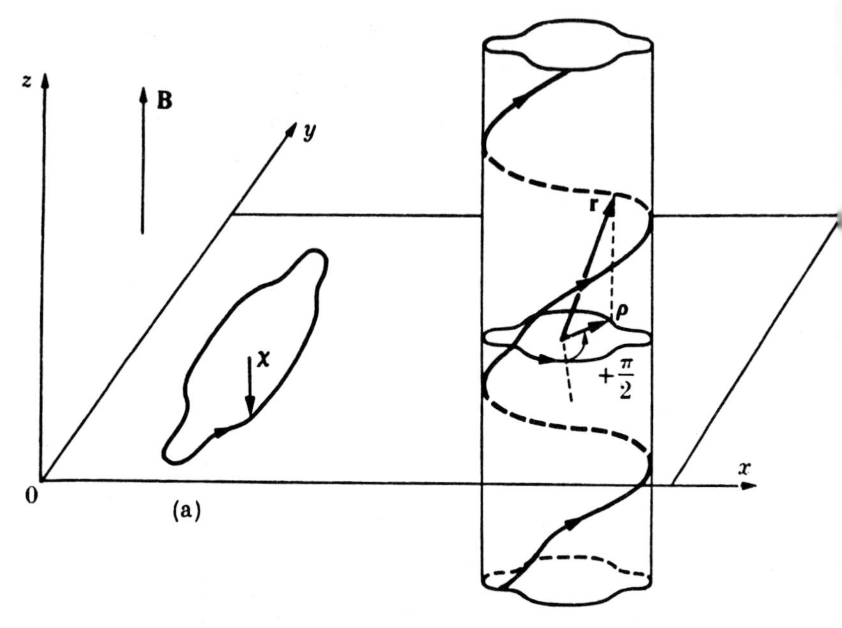

Fig. 9.8. Schematic orbit of an electron in a metal in a magnetic field and projection of orbit onto plane normal to the direction of the magnetic field.

is called the *cyclotron mass*. This projection of the orbit onto a plane normal to $\boldsymbol{B}$ will preserve the same shape if the magnitude of $\boldsymbol{B}$ is altered, but its linear dimensions will be scaled in proportion to $1/\boldsymbol{B}$. It so happens that the detailed shape of this projection of the orbit provides valuable information towards understanding many aspects of the behaviour of the conduction electrons in a metal–not only in a magnetic field but also, perhaps slightly surprisingly, in the absence of a magnetic field too. Ultrasonic methods provide some of the ways of obtaining information about the geometry of the orbit of a conduction electron in a metal in a magnetic field.

If an ultrasonic wave is propagating through a metal it causes an oscillating electric field to be set up in the metal. For a longitudinal (compressional) wave this electric field will arise directly from relative displacements of positive and negative charges, i.e. variations in the electrical charge density, that occur in the compressions and rarefactions. For a transverse wave there are no charge-density variations, but the positive and negative currents associated with the shear waves may not cancel each other; this generates a varying magnetic field and consequently an electric field. The electric field associated with the ultrasonic wave will possess similar wave-like properties to those of the ultrasonic wave

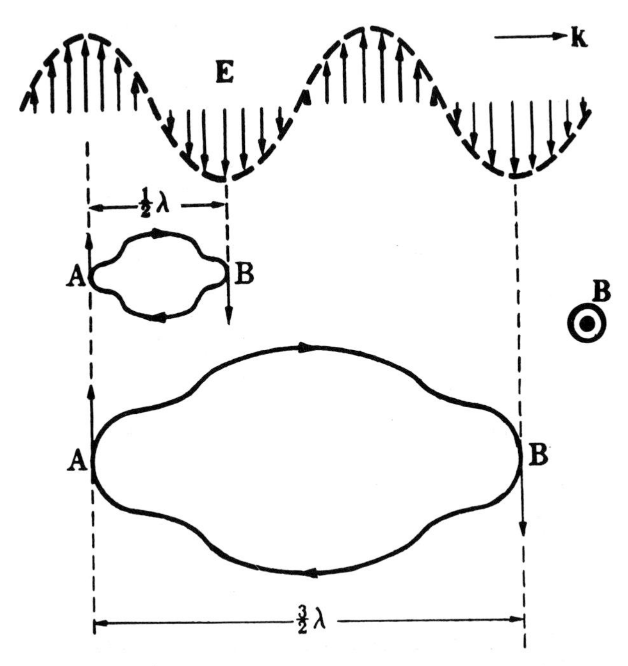

Fig. 9.9. Diagram to illustrate magnetoacoustic geometric oscillations with a transverse ultrasonic wave. ***B*** is normal to the plane of the figure and orbits in real space for two different values of ***B*** are shown. (Adapted from W. Mercouroff, *La Surface de Fermi des Métaux*, Masson, Paris, 1967).

itself. Thus we have what is, in effect, an electric wave which is able to pass through the metal and is not hampered by skin-depth problems as an 'ordinary' electromagnetic wave would be. What is particularly important is that the velocity of this wave is very much slower (typically of the order of $10^3\,m\,s^{-1}$) than the random velocities of the conduction electrons (for which a typical value may be of the order of $10^6\,m\,s^{-1}$). As far as these electrons are concerned the ultrasonic wave appears to produce a sinusoidal electric field which is almost stationary.

Suppose that an ultrasonic wave propagates in the x direction and produces an electric field $\boldsymbol{E}$ in the y direction, where both these directions are normal to $\boldsymbol{B}$, which is in the z direction (see fig. 9.9). Consider two points A and B on opposite sides of the orbit of an electron such that the velocity of the electron when it is at A is exactly opposite to the velocity when the electron is at B. If the distance AB is equal to half an odd integer number of wavelengths of the ultrasonic wave, the electric fields experienced by the electron at A and at B will be exactly anti-parallel. The speed of the electron will, therefore, be increased continuously

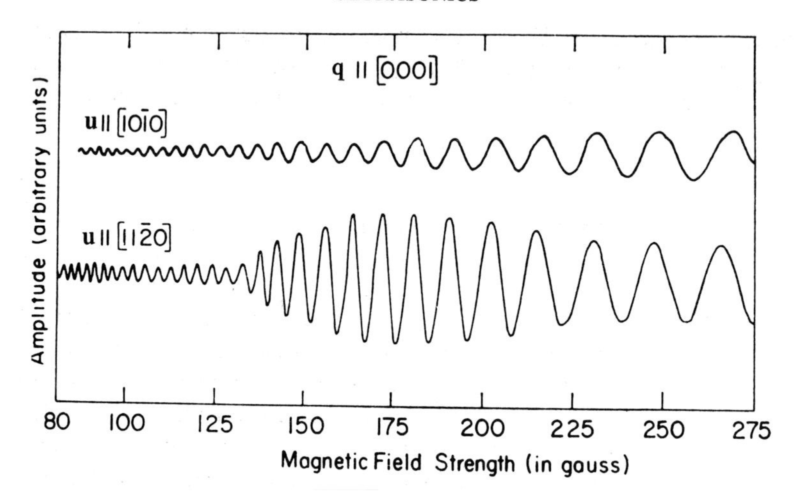

Fig. 9.10. Magnetoacoustic geometric oscillations in Mg. (Adapted from J.B. Ketterson and R.W. Stark, 1967, *Phys. Rev.* **156**, 748.)

as it travels round its orbit. The increase in kinetic energy of the electron will be at the expense of the energy of the ultrasonic wave and the situation is quite closely analogous to the acceleration of an electron, or of some other charged particle, in a cyclotron. The condition for the maximum absorption of energy is therefore

$$AB = (n + \tfrac{1}{2})\lambda_s = (n + \tfrac{1}{2})2\pi/|\mathbf{k}| \tag{9.15}$$

where λ_s is the wavelength and $\mathbf{k}$ is the wave vector of the ultrasonic wave. Since, for a given electron, $AB \propto 1/B$, the resonant absorption will occur as a periodic function in $1/B$ as B is varied. An oscillatory behaviour in the absorption of an ultrasonic wave in a metal, as a function of $1/B$, can indeed be observed (see fig. 9.10). These oscillations are called *magnetoacoustic geometric oscillations* or sometimes, but less appropriately, *magnetoacoustic geometric resonances*. From a detailed study of these oscillations it is possible to determine the dimension AB of the orbit of the electron in the metal.

There are a number of features of magnetoacoustic geometrical oscillations that make the situation more complicated than might be imagined from the above account. First, it is important to realize that each conduction electron in a metal is frequently undergoing collisions that change the magnitude and direction of the velocity of that electron. The average distance that is travelled by an electron between collisions is called the *mean free path* and may be denoted by $\bar{\lambda}$. There will not be any significant resonant absorption of energy from an ultrasonic wave by an electron unless the electron is able to travel several times around its orbit between one collision and the next. That is, for the oscillations in the absorption to occur the mean free path must be substantially larger than AB, i.e.

$$\bar{\lambda} \gg n\lambda_s. \tag{9.16}$$

$\bar{\lambda}$ varies as a function of temperature; for a given metal $\bar{\lambda}$ is smaller at high temperatures and larger at lower temperatures. If one studies the magnitudes of the quantities involved in eqn. (9.16) one finds that the inequality is only satisfied at very low temperatures, say less than about 10 K. Thus one cannot expect to observe the magnetoacoustic geometrical oscillations at room temperature, but must use a considerable amount of specialized cryogenic equipment to perform the experiments. Another point that we did not consider above is the fact that in any metallic specimen there are many different conduction electrons having different sizes of orbit in any given field $\boldsymbol{B}$. Therefore, instead of a single set of oscillations in the ultrasonic attenuation $\alpha(B)$, as B is varied, we might expect that there would be a large number of sets of oscillations, with different periods in $1/B$. Therefore one might have supposed that all these oscillations would smudge each other out. The fact that the oscillations are not smudged out is a result of the fact that only certain special groups of conduction electrons contribute to the oscillations.

It is possible to observe another kind of oscillation in the ultrasonic attenuation coefficient at these low temperatures if the applied magnetic field is sufficiently large. These oscillations arise from changes which occur in the quantization pattern of the conduction electrons as $|\boldsymbol{B}|$ is varied and they are called *magnetoacoustic quantum oscillations*. They are the analogue, for the ultrasonic attenuation, of the quantum oscillations which are known as the *de Haas–van Alphen effect* when they occur in the diamagnetic susceptibility, or as the *Schubnikow–de Haas effect* when they occur in the electrical resistivity. Magnetoacoustic quantum oscillations are also periodic in $1/B$ (see fig. 9.11) and from measurements of the periods of these oscillations it is possible to determine extremal areas of cross section, rather than extremal radii, of the projection of an electron's orbit on a plane normal to $\boldsymbol{B}$. Quantum oscillations can also be observed in the velocity, instead of the attenuation, of ultrasound in a metal; these too are periodic in $1/B$. Both types of magnetoacoustic oscillations, the geometrical oscillations and the quantum oscillations, provide extremely important, and complementary, techniques for studying the behaviour of conduction electrons in metals; they have been widely used in research work on this subject.

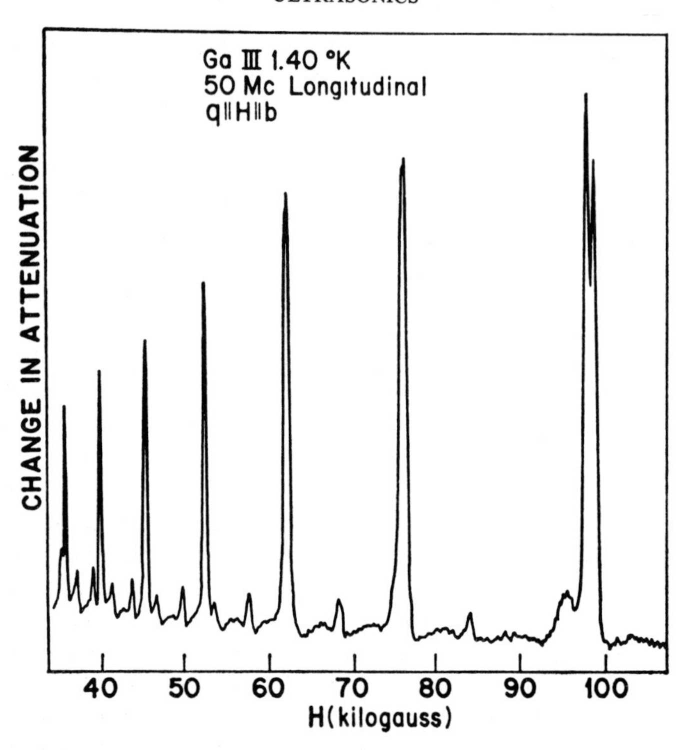

Fig. 9.11. The attenuation of 50 MHz longitudinal ultrasonic waves in Ga as a function of B, showing (giant) quantum oscillations. (From Y. Shapira and B. Lax, 1965, *Phys. Rev.* **138**, A1191.)

(*c*) *The acoustoelectric effect in semiconductors*. If an ultrasonic wave is propagating through a semiconducting material there will be a coupling between the ultrasonic wave and the charge carriers, that is the electrons and holes. Although there is a coupling for both piezoelectric and non-piezoelectric semiconductors, the coupling is stronger and also, perhaps, more easy to understand for piezoelectric semiconductors. In Section 4.2 we discussed piezoelectricity in an insulating crystal. If an ultrasonic wave is propagating in an insulating crystal the local strains produced give rise to local electric fields and so we can regard the ultrasonic wave as being associated with an accompanying electromagnetic wave. If the crystal is a semiconductor rather than an insulator, then it is possible to show that in addition to this accompanying alternating electric field (or electromagnetic wave) there will also be a 'bunching' of the carriers, and consequently a bunching of the space-charge associated with these carriers. We assume that the conductivity of the semiconductor is very small, or that the ultrasonic frequency is very high, so that the carriers do not have time to redistribute themselves to eliminate the alternating 'piezoelectric' fields. If one examines the expression for the electric current density one finds that it consists of two parts. The first is an alternating part that one would expect to find associated with the alternating 'piezoelectric' fields. The second part, however, arises from the interaction between the 'bunched' space charges and the oscillating 'piezoelectric' fields; this part is a dc component and is called the *acoustoelectric current*. The existence of this current is known as the *acoustoelectric effect*.

Ultrasonic waves travelling in a piezoelectric material which is also semiconducting will experience both dispersion (velocity depending on frequency) and attenuation (amplitude decreasing with distance travelled) due to their interactions with the carriers. The attenuation occurs because energy is transferred from the ultrasonic waves to the carriers. An ultrasonic wave carries momentum as well as energy and so, in addition to losing energy, the ultrasonic waves also lose momentum as they travel through a medium. In most circumstances, when an ultrasonic wave is attenuated, the momentum that it loses is transferred to the whole specimen of the material. Recalling that momentum = mass × velocity and using typical values for the momentum of an ultrasonic wave and for the mass of a specimen, we would find that the velocity which the specimen would acquire would be imperceptibly small. However, in the case of the acoustoelectric effect the momentum which is lost, or at least a considerable proportion of it, is transferred to the carriers of electric charge, that is to the electrons and holes instead of to the entire material. Because the mass of an electron, or of a hole, is very small the momentum that the carriers acquire is large enough to increase their velocities quite appreciably. In other words the acoustoelectric current can be regarded as arising from the transfer of momentum from the ultrasonic waves to the carriers. The acoustoelectric effect was first observed experimentally by Weinreich and White in 1957 in *n*-type germanium using transverse ultrasonic waves with a frequency of about 60 MHz.

The inverse of the acoustoelectric effect, namely the appearance of high-frequency current oscillations when a suitable specimen is subjected to a high electric field, was observed by Gunn in 1963 using *n*-type gallium arsenide and indium phosphide. This is known as the *Gunn effect*. The discovery of the Gunn effect aroused great interest because of the possibility that it raised for the development of simple solid-state sources of microwave power.

9.4. *Ultrasonic bubble chambers*

The high frequency of ultrasonic vibrations can be exploited in the construction of bubble chambers which achieve a higher repetition rate than can be obtained with more conventional bubble chambers using mechanical means for producing compressions and rarefactions.

A bubble chamber contains an enclosed volume of liquid at a pressure that is lower that its svp (saturated vapour pressure). As an ionizing particle passes through the liquid and loses energy, it will tend to evaporate the liquid to form small bubbles in the track of the particle. These bubbles will tend to collapse as a result of surface tension. For any given liquid there is a certain pressure, p_s, which is known as the 'sensitive limit'. p_s depends on the internal pressure due to surface tension in the largest "microbubble" (typically of diameter $\sim 10^{-6}$ cm) which can be created instantaneously with energy deposited by the ionizing particle. If the pressure of the liquid is less than p_s the microbubbles produced by the passage of an ionizing particle will not be collapsed but will grow large enough to be visible (say diameter $\sim 10^{-2}$ cm) in a very short time (less than 1 ms) and the track of the particle, as manifested by the line of bubbles, can be photographed. The bubble chamber can be 'wiped clean' again, in readiness for producing and photographing the next set of tracks, by increasing the pressure again so that it exceeds p_s. The repetition rate that can be achieved in a conventional bubble chamber with a mechanical piston is very low, say $\leqslant 20$ Hz. In an ultrasonic bubble chamber there is no piston and the expansion of the liquid to a pressure below p_s is achieved in the expansion half of the cycle of an ultrasonic wave. If the conditions are right, a bubble which is formed and grows in the expansion half of a cycle may not be collapsed completely in the subsequent compressional half cycle and will then be able to achieve a net growth over a succession of cycles. It would be possible to use either travelling or stationary ultrasonic waves, but in practice it is found to be better to use stationary waves. While it is now established that ultrasonic bubble chambers can be made to work, they are still very much at the development stage, rather than being routine or standard equipment.

9.5. *Chemical effects of ultrasound*

It is a reasonably familiar everyday experience that we may be able to accelerate a chemical reaction involving liquids or solutions by stirring the materials or by shaking the vessel containing them. In a rather naïve way we can, therefore, see that it is not surprising that the 'agitation' produced by ultrasound in a liquid may 'catalyse' or initiate chemical reactions. In nearly all ultrasonic cleaning applications (Section 8.2) one uses an ultrasonic intensity that exceeds the threshold intensity for cavitation in the liquid. So also in the case of chemical reactions, it has been found by experiment that in most cases significant effects

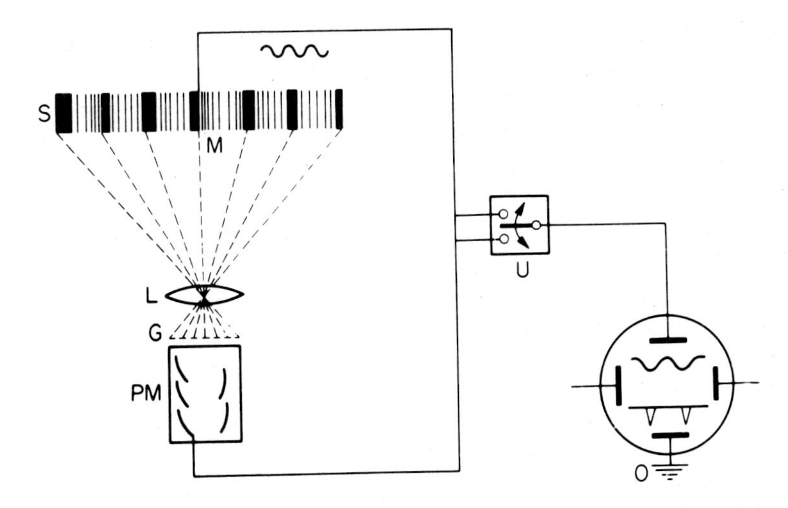

Fig. 9.12. Schematic arrangement for the simultaneous measurement of the alternating acoustic pressure and the moments of appearance of sonoluminescent spikes. (From K. Negishi, 1960, *Acustica,* **10**, 124.)

of the ultrasound will only occur if the threshold intensity for cavitation is exceeded.

Many chemical reactions that occur in an ultrasonic field are accompanied by an emission of light and this phenomenon is known as *sonoluminescence*. Thus luminescence may indicate that a chemical reaction has been initiated by ultrasound. It does not, however, follow that all chemical reactions initiated by ultrasound involve the emission of light. The instant when luminescence appears is now commonly held to correspond to the beginning of a chemical reaction at the place where a bubble collapses. Therefore an important insight into the mechanism of ultrasonic chemical reactions can be obtained by relating the time at which luminescence occurs to the cycle of local pressure variations in the ultrasonic wave. In fig. 9.12 the microphone (M) records the changes in the local pressure in the ultrasonic field, while the lens (L) and photomultiplier (PM) are used to detect the light emitted. Some results, in which there are two 'spikes' of sonoluminescence, are shown in fig. 9.13.

Many theories have been proposed to try to explain the chemical consequences of cavitation produced by ultrasound. A recent article by Śliwiński (see Further Reading), for example, outlines about half a dozen different theories.

We shall now mention a small selection of chemical reactions that occur in an ultrasonic field; most of them involve aqueous solutions. It has been suggested that free radicals of H and OH are produced in the

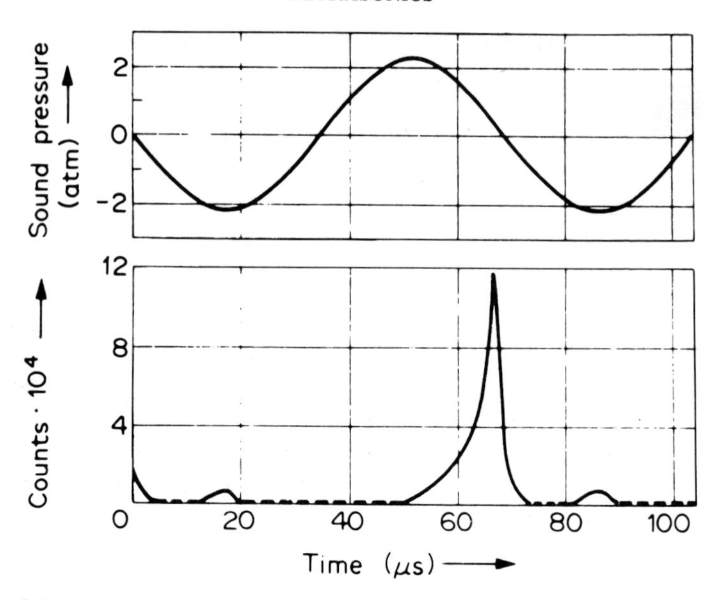

Fig. 9.13 Distribution of sonoluminescence intensity (lower figure) throughout a period of acoustic pressure (From P.D. Jarman and K.J. Taylor, 1970, *Acustica,* 23, 243.)

vapour phase within the cavitation bubbles and there is some experimental evidence for this. These free radicals may react among themselves in various ways; for example

$$\begin{aligned}
&OH + OH \rightarrow H_2O_2 \\
&H + H \rightarrow H_2 \\
&OH + OH \rightarrow H_2O + O \\
&2O \rightarrow O_2 \\
&H_2O_2 + OH \rightarrow H_2O + HO_2 \\
&HO_2 + HO_2 \rightarrow H_2O_2 + O_2 \\
&HO_2 + H \rightarrow H_2O_2 \\
&H_2O_2 + HO_2 \rightarrow H_2O + O_2 + OH \quad \text{etc.}
\end{aligned} \tag{9.17}$$

The occurrence of the radicals H, OH and HO_2 and of H_2O_2 has been observed experimentally. If the water contains other dissolved substances and the molecules of these substances enter the cavitation bubble, then they can react with the disintegration products of the water. For example, if the water in an ultrasonic field contains dissolved oxygen the production of hydrogen peroxide, H_2O_2, is greatly enhanced. Thus:

$$\begin{aligned} &H^+ + O_2 \rightarrow HO_2 + 2e \\ &HO_2 + HO_2 \rightarrow H_2O_2 + O_2 \\ &O_2 \rightarrow O + O \\ &O_2 + O \rightarrow O_3 \\ &H_2O + O_3 \rightarrow H_2O_2 + O_2 \quad \text{etc.} \end{aligned} \tag{9.18}$$

Thus we can regard a solution of oxygen in water subjected to ultrasound as an oxidizing agent. Similarly, a solution of hydrogen in water subjected to ultrasound, can be regarded as a reducing agent; for example Fe(III) ions are reduced to Fe(II) ions in this way.

The application of ultrasound to other aqueous solutions produces a number of interesting results. Thus aqueous solutions of oxygen and nitrogen produce nitric oxide, NO:

$$\begin{aligned} &N_2 \rightarrow N + N \\ &O_2 \rightarrow O + O \\ &N + O \rightarrow NO \end{aligned} \tag{9.19}$$

solutions of nitrogen and hydrogen produce ammonia:

$$\begin{aligned} &N_2 \rightarrow N + N \\ &H_2 \rightarrow H + H \\ &N + H \rightarrow NH \\ &NH + H_2 \rightarrow NH_3 \end{aligned} \tag{9.20}$$

solutions of carbon monoxide and hydrogen produce formaldehyde:

$$\begin{aligned} &CO \rightarrow CO^+ + e \\ &CO^+ + H_2 + e \rightarrow HCHO \end{aligned} \tag{9.21}$$

and solutions of nitrogen and methane, or of nitrogen, carbon monoxide and hydrogen produce hydrogen cyanide HCN. Another interesting fact that might not, at first sight, be expected is that various sonochemical reactions may be accelerated by having an inert gas dissolved in the water. Thus, for example, the synthesis of H_2O_2 in water saturated with oxygen occurs more easily if argon or helium is also dissolved in the water. The influence of ultrasound on aqueous solutions of numerous organic molecules has also been investigated. The important feature of the application of ultrasound on many organic compounds seems to be that it disrupts

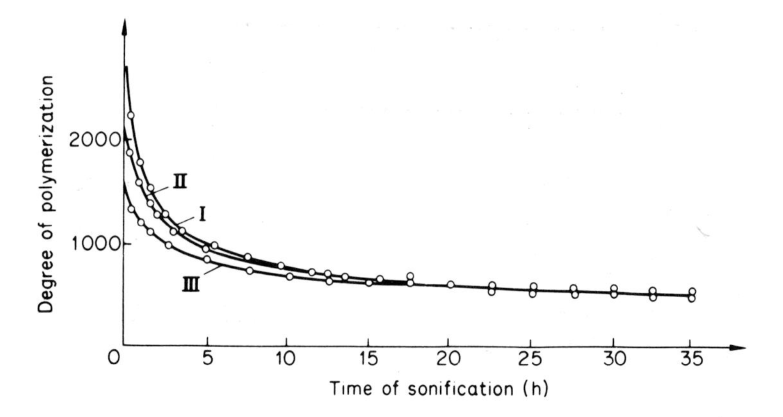

Fig. 9.14. Effect of depolymerization of polystyrene, with various initial degrees of polymerization, dissolved in benzene versus the sonification time; ultrasonic frequency, 740 kHz, intensity 12 W cm^{-2}). (From M.A.K. Mostafa, 1958, *J. Polymer Sci.*, **33**, 311.)

chemical bonds and activates molecules that are resistant to other agents. Thus, for example, the benzene ring which is highly resistant to attack by other agents can be disrupted by ultrasonic action. The effect of ultrasound on organic molecules has generally been investigated in aqueous solutions but, occasionally, organic solvents have been used too. As with small organic molecules, so also in a similar manner the principal effect of ultrasound on very large molecules or on polymer chains is to break the large molecules or polymer chains into smaller ones (see fig. 9.14).

The effect of ultrasound on electrochemical reactions has also been studied quite widely. A schematic illustration of a system for studying the effect of ultrasound on an electrochemical reaction is shown in fig. 9.15. By using two electrolytic baths in series we can be sure that the same current passes through each. This enables a rather direct determination of the effect of ultrasound to be made by comparison between the results obtained in the two baths. We are concerned with the effect of ultrasound on the electrolytic system as a whole, that is on processes at the electrodes, on processes in the electrolytic solution, and on the electrokinetic processes in the solution-electrode layer. The overall effect of the application of ultrasound is to improve the efficiency of electrolysis and to facilitate processes of electrolytic oxidation and reduction. In the electrolytic deposition of metals, not only does the application of ultrasound improve the efficiency of the process, but also the layers of metal deposited are of a higher quality. Moreover, the properties of these layers can be controlled to some extent by varying the frequency and the intensity of the ultrasound that is used.

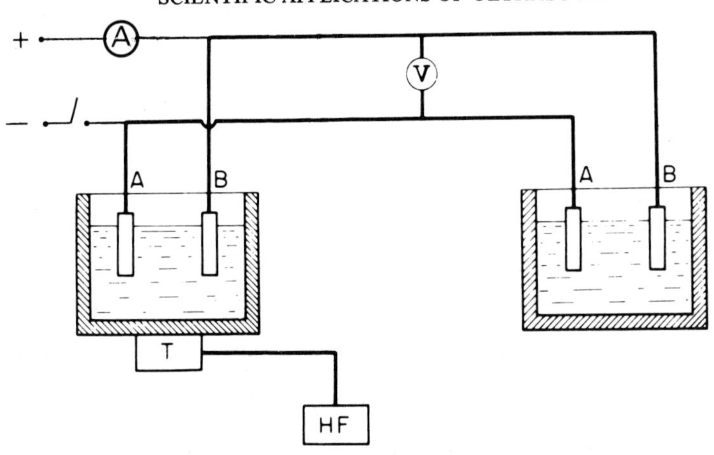

Fig. 9.15. A schematic set-up for examination of electrolytic processes in an ultrasonic field; (a) irradiated electrolytic bath, (b) unirradiated electrolytic bath. T = ultrasonic transducer, HF = generator.

In Section 3.2 we mentioned thermal relaxation and structural relaxation as mechanisms that contribute to the attenuation of ultrasound. α_{relax}, the part of the attenuation coefficient due to these relaxation mechanisms can be determined by using eqn. (3.18)

$$\alpha = \alpha_{vis} + \alpha_{th} + \alpha_{relax} \tag{3.18}$$

where α is the total measured attenuation coefficient, and α_{vis} and α_{th} can be calculated. Thus we have the possibility of ultrasonic spectroscopy, that is measurements of ultrasonic attenuation or less commonly ultrasonic velocity, as a tool in chemical research on the structures and properties of molecules, principally organic molecules and polymers.

It would seem to be a fair summary of the present state of the subject of sonochemistry to say that it is known that cavitation produces chemical effects, that a considerable amount of knowledge has been gained experimentally about these effects, that they can usefully be exploited, and that many theories have been proposed to explain them, but that much work still needs to be done to discriminate among the available theories.

10. Safety

Apart from one or two passing remarks we have not mentioned the question of safety so far in this book. We have seen in the later chapters that ultrasound is now widely used in an ever-increasing range of applications, principally in industry, in medicine, and in scientific research. It is, therefore, very important to consider the question of the safety of the operators of ultrasonic equipment, the safety of patients and the safety of the general public when processes involving ultrasound are being used. The safety of the general public is not a serious problem. Ultrasonic energy is not going to harm the general public by escaping from a factory or hospital through the windows or up the chimney. Neither is there a serious effluent problem, for example the solvents used in ultrasonic cleaning processes are certainly no more noxious, and are probably less noxious, than those used in non-ultrasonic cleaning processes. Nor does a product, be it anything from salad cream to a plastic toy, which has been treated by an ultrasonic industrial process contain residual ultrasonic energy to harm the subsequent user of the product. These may seem rather trivial points, but nevertheless they are important in these days when people are very conscious of pollution problems.

The case of the remaining people, namely patients and the operators of industrial, medical or scientific ultrasonic equipment, needs to be considered more carefully. The main factors which have to be considered are

(*a*) the intensity of the ultrasound generated
(*b*) the frequency, and
(*c*) the dose received by the patient or operator.

The importance of the intensity should be obvious; the more energy is generated, then the more energy there may be available to harm the operator or a patient. The frequency is probably only indirectly important in the sense that if we are considering airborne ultrasound from a piece of ultrasonic equipment, the frequency will affect the attenuation experienced by any ultrasound that escapes from the equipment. Since the attenuation is proportional to ν^2 (see Section 3.2) we can see that the higher the frequency that is used, the greater will be the attenuation experienced by that ultrasound in the air and, correspondingly, the lower is the danger to the operator of the equipment. For a given intensity of generated ultrasound, the dose received by a patient who is directly coupled to the ultrasonic source will be enormously greater than the dose received, by transmission through the air, by anyone else in the vicinity. The reasons for this are two-fold; first there is the very large acoustic mismatch between the ultrasonic generator and the air and also between the air and the human body and, secondly, there is the rather high attenuation of ultrasound when it is travelling through the air.

In medical applications where the patient is coupled directly, or via a surgical instrument, to the ultrasonic generator, it would be extremely naïve to assume that there was no danger to the patient. Experiments have been performed on animals such as mice and rabbits to study the possibility of damage resulting from the application of ultrasound to various different tissues and organs. By using sufficiently high doses of ultrasonic energy it is possible to produce permanent damage to an organ and ultimately, at very high doses, the death of the animal. For each organ it is possible to determine an applied intensity/irradiation duration threshold curve (fig. 10.1). For each curve a pair of values of intensity and duration corresponding to a point below a curve in fig. 10.1 there is no damage. Thus a point on the curve represents either the minimum intensity for a given duration or the minimum duration for a fixed intensity to cause damage. The exposures typically applied using commercial instruments are also indicated in fig. 10.1. Thus it would appear that *as far as current diagnostic applications are concerned* the doses used are relatively low and ultrasound is a good deal safer than other available diagnostic aids such as X-rays (see Section 5.3). However, safety standards in this area have not so far been established and, in the meantime, it is wise not to subject patients to more exposure to ultrasound than is absolutely necessary. In surgical applications (see Section 8.5) the intensities used are likely to be larger than those used in diagnosis or, at any rate, the ultrasound will be coupled to potentially dangerous instruments such as scalpels, etc. There is, obviously, some potential danger to the patient, in the sense that the ultrasound is involved as an agent in causing permanent changes to cells and there is the possibility of damage to cells other than those that it is intended to affect. Thus the surgeon's hand may slip, but this is a risk that is already known and accepted in non-ultrasonically-assisted surgery. By the standards of modern medical and surgical practice the side effects of the use of ultrasound are, at present, not thought to be unacceptably high.

In considering the operators of ultrasonic equipment it is the possible danger from airborne ultrasound, or associated airborne audible sound, that is likely to be of primary importance. Those who work in scientific research

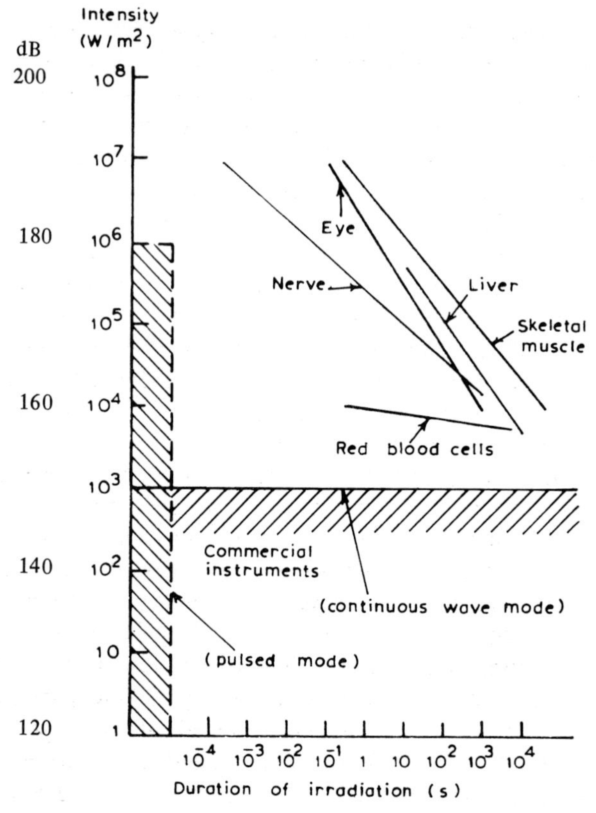

Fig. 10.1. Intensity/duration thresholds for some mammalian organs (From P.D. Edmonds, *Interaction of Ultrasound and Biological Tissues*, U.S. D.H.E.W. Publication (FDA) 73-8008, page 299, 1972.)

laboratories are, or should be, generally aware of the potentially hazardous nature of many of the operations or pieces of equipment contained in the laboratory. Although it would be foolish to ignore them completely, the hazards to personnel arising *directly* from the use of ultrasound in scientific experiments are probably rather low, compared with other laboratory hazards. There are two reasons for this. First the intensities used are generally not very great and secondly the frequencies are likely to be very high and so any emergent ultrasound will be heavily attenuated in the air. In medical diagnostic applications (see Section 5.3) the same reasons would suggest that the hazard from airborne ultrasound to the operators of ultrasonic equipment was also rather low.

In the application of ultrasound to industrial processes the intensities used are generally higher than in control or diagnostic applications and in research work. Thus there is simply more energy available as a potential source of harm to operators of industrial ultrasonic equipment. In addition to this, the ultrasonic frequencies that are used are generally rather low, say in the range of 20–40 kHz. However, although the main frequency being used is above the human threshold frequency of hearing, it is quite probable that there will be some associated generation of, or conversion to, audible sound. Moreover, ultrasound of the basic operating frequency of the equipment, and which has not been converted to lower frequencies, will be only in the kHz range rather than the MHz range; consequently if some of this ultrasound escapes into the air it will be less strongly attenuated than would ultrasound in the MHz range. Therefore on both counts, namely high intensity and rather low frequency, the industrial use of ultrasound as a form of energy should not automatically be regarded as free of hazard but should be examined rather carefully.

We may divide the consideration of industrial airborne ultrasound into auditory effects, physiological effects and subjective effects. For the auditory effects, that is the effects of airborne ultrasound on hearing at audible frequencies, the available evidence from ultrasonic tests seems to indicate that no temporary or permanent reduction in the threshold frequency of hearing for audible sound is caused by exposure to typical intensities of airborne ultrasound in industrial ultrasonic applications. Figure 10.2 summarizes the available knowledge of the physiological effects of exposure to very high intensity ultrasound based on experimental results and some extrapolation for humans. It should be emphasized that, for a number of reasons, care should be exercised in extrapolating the effects on small animals to the case of humans. First, the impedance mismatch between skin tissue and the air is very great but for a fur-covered animal the fur acts as an impedance matching device; hence body-temperature rise in haired mice occurs at a lower exposure level of ultrasound than it does for hairless mice (see fig. 10.2). Secondly, the ratio of surface area to body mass is much greater for small animals than for man and so these animals have more difficulty than man in dissipating the heat generated by airborne ultrasound. Thirdly, the lower ultrasonic frequencies may be audible to these animals (see fig. 1.1). Figure 10.2 can be considered as an extension of Table 3.2 to higher intensity levels. It should also be appreciated that fig. 10.2 refers to 'whole-body' exposure to (airborne) ultrasound, whereas fig. 10.1 refers to the direct exposure of an individual organ. A recommended limit of 100–110 dB for industrial exposure would clearly be well below the levels at which, according to fig. 10.2, any physiological effects of the ultrasound will occur. To that extent therefore such processes involving exposure up to that limit can be considered safe. Notice that 110 dB is much higher than the recommended

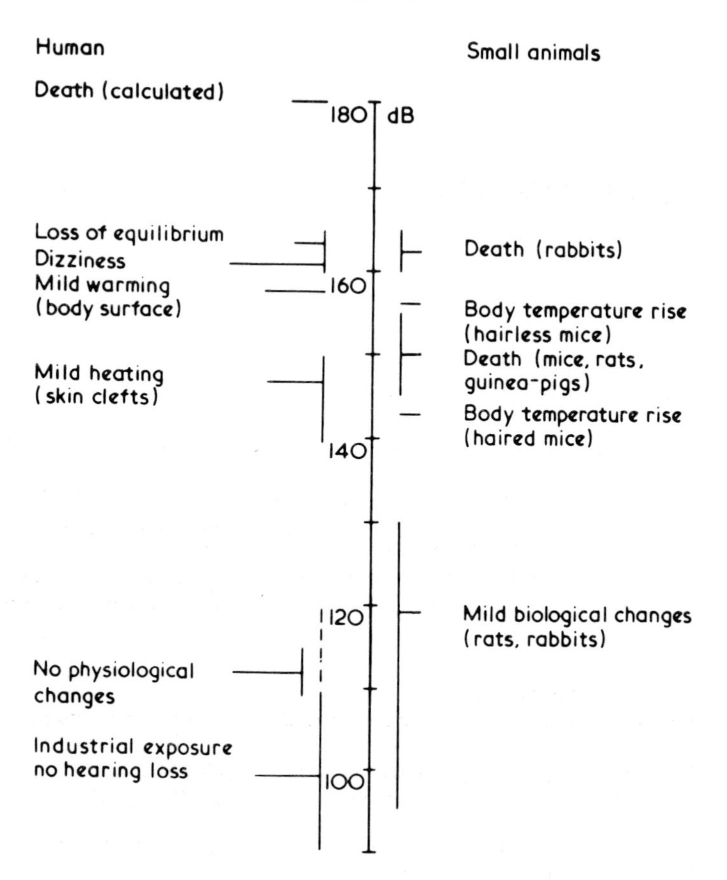

Fig. 10.2. Physiological effects of ultrasound. (From W.I. Acton, 1974, *Ultrasonics*, **12**, 124.)

maximum noise levels in Table 3.2. The subjective effects, such as fatigue, headaches, nausea, etc., which may be experienced by the operators of industrial ultrasonic equipment, and which are not necessarily psychosomatic, are now thought to be due to high-frequency *audible* sound emitted or generated by the ultrasonic equipment, rather than to ultrasound itself. Thus for very low ultrasonic frequencies, which may be audible to some operators of industrial ultrasonic equipment, the recommended limit of about 75 dB in Table 3.2 (which applied to audible sound) should be applied rather than a limit in the vicinity of 100–110 dB.

Further reading

General references

J. Blitz, *Fundamentals of Ultrasonics* 2nd ed., (Butterworth, London, 1967).

G.L. Gooberman, *Ultrasonics, Theory and Applications* (English Universities Press, London, 1968).

There is a whole series of volumes, published during about the last ten years, entitled *Physical Acoustics, Principles and Methods*, edited by W.P. Mason (later W.P. Mason and R.N. Thurston) and published by Academic Press, New York. Many useful and readable articles will be found in the journal *Ultrasonics*, published by IPC Science and Technology Press, London.

Some older general books

L. Bergmann, *Ultrasonics and Their Scientific and Technical Applications* (Wiley, New York, 1943).

P. Vigoureux, *Ultrasonics* (Chapman and Hall, London, 1950).

Parts of A.B. Wood, *A Textbook of Sound* (Bell, London, 1955).

These references may be supplemented, where desired, for certain particular chapters or sections.

Chapter 1

R.W.B. Stephens, 'An historical review of ultrasonics' in *Ultrasonics International 1975 Conference Proceedings* (IPC Science and Technology Press, London, 1975) p. 9.

Section 1.3

R.T. Beyer, 1950, Radiation pressure in a sound wave, *Amer. J. Phys.*, **18**, 25.

A.P. French, *Vibrations and Waves* (Nelson, London, 1971) pp. 243-4.

Chapter 2

Standard texts on 'sound' or on 'vibrations and waves' are very useful here.

Section 2.3

W.M. Wright, 1969, Experiments for the study of acoustic wave phenomena, *Amer. J. Phys.*, **37**, 110.

Section 2.4

J.F. Nye, *Physical Properties of Crystals* (Clarendon Press, Oxford, 1957) chapter 8.

W.A. Wooster, *Tensors and Group Theory for the Physical Properties of Crystals* (Clarendon Press, Oxford, 1973) chapters 7 and 8.

Section 3.3

A. Śliwiński and A. Walasiak, 1967, Ultrasonic light diffraction in automatic control of concentration in binary solutions, *Ultrasonics,* **5**, 163.

Section 4.1. Early work on whistles and sensitive flames

F. Galton, *Inquiries into Human Faculty and Its Development* (Macmillan, London, 1883).

J.W. Strutt, Baron Rayleigh, *The Theory of Sound*, vol. 1, 1877, 2nd ed., 1894, vol. 2, 1878, 2nd ed., 1896 (Macmillan, London, reprinted by Dover Books, New York, 1945).

Section 4.2

W.P. Mason, *Piezoelectric Crystals and Their Application to Ultrasonics*, D. Van Nostrand, Princeton, New Jersey, 1950).

J.F. Nye, *Physical Properties of Crystals* (Clarendon Press, Oxford, 1957) chapter 7.

W.A. Wooster, *Tensors and Group Theory for the Physical Properties of Crystals* (Clarendon Press, Oxford, 1973) chapters 4 and 5.

Section 4.3

M.P. Felix, 1974, Laser generated ultrasonic beams, *Rev. Sci. Instrum.*, **45**, 1106.

Section 5.1

S.B. Palmer and G.A. Forster, 1970, Experiments on pulsed ultrasonics, *Amer. J. Phys.*, **38**, 814.

S. Chapman, 1971, Size, shape and orientation of sonar targets measured remotely, *Amer. J. Phys.*, **39**, 1181.

D. Cushing, *The Detection of Fish* (Pergamon, Oxford, 1973).

Section 5.2

J. Krautkrämer and H. Krautkrämer, *Ultrasonic Testing of Materials* (Springer, Berlin, 1969).

R.P. Mackal, *The Monsters of Loch Ness* (Macdonald and Janes, London, 1976).

Section 5.3

P.N.T. Wells, *Physical Principles of Ultrasonic Diagnosis* (Academic Press, London, 1969).

L. Filipczynski, 'Ultrasonic medical diagnostic methods', in *Acoustic 1974, the Invited Lectures Presented at the Eighth International Congress on Acoustics, London, 1974,* R.W.B. Stephens (ed.) (Chapman and Hall, London, 1975).

J. Vaný́sek, J. Preisová, and J. Obraz, *Ultrasonography in Ophthalmology* (Butterworths, London, 1970).
J. François and F. Goes (eds.), *Ultrasonography in Ophthalmology* (S. Karger, Basel, 1975).
S.N. Hassani, *Real Time Ophthalmic Ultrasonography* (Springer, New York, 1978).

Chapter 6
D.R. Griffin, *Listening in the Dark; The Acoustic Orientation of Bats and Men* (Yale University Press, New Haven, 1958).
G. Sales and D. Pye, *Ultrasonic Communication by Animals* (Chapman and Hall, London, 1974).

Section 6.2
G.W. Pierce, *The Songs of Insects* (Harvard University Press, Cambridge, Mass., 1948).
K.D. Roeder, *Nerve Cells and Insect Behavior* (Harvard University Press, Cambridge, Mass, 1963, revised 1967).

Section 6.3
W.N. Kellogg, *Porpoises and Sonar* (Chicago University Press, Chicago, 1961).

Section 6.4
L.L. Clark (ed.), *Proceedings of the International Congress on Technology and Blindness* (The American Foundation for the Blind, New York, 1963).

Chapter 7
G. Wade (ed.), *Acoustic Imaging: Cameras, Microscopes, Phased Arrays and Holographic Systems* (Plenum, New York, 1976).

Section 7.1
P. Kang and F.C. Young, 1972, Diffraction of laser light by ultrasound in liquid, *Amer. J. Phys.*, **40**, 697.
A.G. Hellier, S.B. Palmer and D.G. Whitehead, 1975, An integrated circuit pulse-echo overlap facility for measurement of velocity of sound, *J. Phys. E.; Scientific Instrum.*, 8, 352.

Section 7.2
J. Szilard and M. Kidger, 1976, A new ultrasonic lens, *Ultrasonics*, **14**, 268.
C.H. Jones and G.A. Gilmour, 1976, Sonic cameras, *J. Acoustic Soc. Amer.*, **59**, 74.

Section 7.3 Resolution
G. Sales and D. Pye, *Ultrasonic Communication by Animals* (Chapman and Hall, London, 1974) appendix.
S. Chapman, 1971, Size, shape and orientation of sonar targets measured remotely, *Amer. J. Phys.*, **39**, 1181.

Section 7.5

A.F. Metherell and L. Laramore (eds.), *Acoustical Holography*, vol. 2 (Plenum, New York. 1970).

E.E. Aldridge, *Acoustical Holography* (Merrow, Watford, 1971).

Section 7.6

A.A. Pollock, Acoustic emission, in *Acoustics and Vibration Progress*, vol. 1, R.W.B. Stephens and H.G. Leventhall (eds.) (Chapman and Hall, London, 1974) chapter 2.

Chapter 8

B. Brown and J.E. Goodman, *High Intensity Ultrasonics* (Iliffe, London, 1965).

Section 8.5

I.P. Goliamina, Ultrasonic surgery, in *Acoustics 1974, The Invited Lectures Presented at the Eighth International Congress on Acoustics, London, 1974*, R.W.B. Stephens (ed.) (Chapman and Hall, London 1975) p. 63.

Sections 9.1–9.3

R. Truell, C. Elbaum and B.B. Chick, *Ultrasonic Methods in Solid State Physics* (Academic Press, New York, 1969).

Section 9.3

A.P. Cracknell and K.C. Wong, *The Fermi Surface: Its Concept, Determination, and Use in the Physics of Metals* (Oxford University Press, Oxford, 1973).

A.W.B. Taylor, *Superconductivity* (Wykeham Publications, London, 1970).

J.W. Tucker and V.W. Rampton, *Microwave Ultrasonics in Solid State Physics* (North-Holland, Amsterdam, 1972).

Section 9.4

R.C.A. Brown and H.J. Hilke, 1972, The development of ultrasonic bubble chambers, *Physics Bulletin,* **23**, 215.

Section 9.5

A. Śliwiński, Chemical aspects of ultrasonics, in *Acoustics and Vibration Progress,* vol. 1, R.W.B. Stephens and H.G. Leventhall (eds.) (Chapman and Hall, London, 1974) chapter 3.

Chapter 10

I.E. El'Piner, *Ultrasound: Physical and Biological Effects* (Consultants Bureau, New York, 1964).

P.N.T. Wells, *Physical Principles of Ultrasonic Diagnosis* (Academic Press, London 1969) chapter 7.

W.I. Acton, 1974, The effects of industrial airborne ultrasound on humans, *Ultrasonics,* **12**, 124.

W.I. Acton, 1975, Exposure criteria for industrial ultrasound, *Ann. Occup. Hyg.,* **18**, 267.

M. Hussey, *Diagnostic Ultrasound: An Introduction to the Interactions between Ultrasound and Biological Tissues,* (Blackie, Glasgow, 1975).

Index